The Elements of Technical Writing

HARCOURT BRACE JOVANOVICH, INC.
New York/San Diego/Chicago/San Francisco/Atlanta

The Elements of Technical Writing

Joseph A. Alvarez SANTA ROSA JUNIOR COLLEGE

ISBN: 0-15-522160-4
Library of Congress Catalog Card Number: 79-3293
Printed in the United States of America

COVER Detail of Leonardo da Vinci's Spinning Wheel c. 1490. Reproduced by permission of Biblioteca Ambrosiana, Milano.

TO KRISTA

. . . IT IS AS CRAFTSMEN THAT WE
GET OUR SATISFACTION AND OUR PAY.

—Learned Hand

The Shakespearean scholar G. B. Harrison has written that his most effective training in writing was as a staff captain in World War II coordinating 72 miscellaneous military units. "It is far easier to discuss Hamlet's complexes," he explained, "than to write orders which ensure that 5 working parties from 5 different units arrive at the right place at the right time equipped with proper tools for the job."

Captain Harrison faced what every technical writer faces: the necessity of being clear and accurate, coherent and concise. This challenge, and how to meet it, is the subject of this book.

Technical writing is coming of age, and the following pages are directed to anyone interested in it. This includes technical and business students as well as writers in business, industry, the sciences, and the professions. But since the book's premise is that technical writing is chiefly *writing* and only secondarily technical, it is addressed to anyone who writes in school or on the job.

The emphasis on technical writing as *writing* may be more radical than it seems. Many texts treat technical writing as a separate genre, a distinctive category of writing, with "psychological," "linguistic," "sequential," and "functional" differences that distinguish it from other writing. But what distinguishes technical writing from other writing is none of these; rather, it is special knowledge, and often a special vocabulary. The authors of these texts perhaps confuse the sometimes esoteric nature of technical material with the straightforward description of it. They may have forgotten that writing—any kind of writing—remains the selection and arrangement of words, whatever those words may be, whatever purpose they may have, whatever subject they may refer to.

Another principle of this book is that usage is the arbiter of language, that rules should reflect usage rather than dictate it. In that sense, this book is descriptive rather than prescriptive. I also believe that the standard of writing—especially of technical writing—should be less "correctness" and more effectiveness: does it do the job? The questions then become: how does it work and why? If it doesn't, why not? And what are some alternatives?

In matters of form, I have followed the University of Chicago Press' Manual of Style; but I have not adhered to it rigidly. I believe that function ultimately determines form, and when the two conflict, I favor function. In matters of form, there is seldom absolute agreement anyway. Like universal language, universal form eludes us; most disciplines, most companies, create their own. And the value of one or another lies less in any inherent virtue than in the force of habit and the security of familiarity.

The book itself is designed not only as a guide to technical writing but as a reference for writers. It consists of 50 axioms arranged in 8 sections that cover major problem areas in writing: Principles, Words, Sentences, Organization, Punctuation, Format, Style, and Method. In science an axiom is, according to Webster's Third New International Dictionary, "a proposition regarded as a self-evident truth." Writing, however, is not a scientific activity. My use of axiom rests on the meanings of its original Greek roots: axioun, "to think worthy," and agein, "to weigh." In writing there are opinions, not truths, and their worth depends on the perspective of the writer and the reader.

The examples in the text follow the points they illustrate. I have drawn their subject matter from science, business, industry, philosophy, politics, education, law, and even poetry. To save space, I have used portions of the text itself as examples. I also have suggested several exercises at the end of each section. My first priority in the examples and exercises has been to clarify the point; my second has been to make them interesting.

My purpose here is to help writers learn how to say what they mean and to say it clearly. In the end, writing comes down to a succession of choices that the writer must resolve: this idea here, that one there; one word instead of another; one long sentence rather than two short ones; and so on. Basically, the book is designed to help writers make informed choices.

One of my own choices was to make the book itself selective. Rather than cover everything in the field, I have focused on the essential elements of technical writing and those principles that apply to the significant problems of writing. To do otherwise would have produced a massive volume that would have smothered the essential points. For example, instead of listing the countless variations in the format and content of technical reports, I looked for basic principles that could be applied broadly. Nor did I include material on oral reports, statistical analysis, or other marginal subjects. All but two of the axioms refer to writing, and those two refer to graphics and research, vital elements in technical writing. In short, the emphasis here is on the process of writing: how to produce an effective report or paper.

Joseph A. Alvarez

Acknowledgements

A piece of writing springs from more roots than can be traced. This book is no exception. It has been shaped by my own education, by my experience as a professional writer and editor and writing instructor, and by the people I have known either personally or professionally. The people are too numerous to name individually, but I wish to single out one, Angus Cameron.

I also wish to thank my colleagues in the English and various technical departments of Santa Rosa Junior College; the staff of the College's Plover Library; and my editors at Harcourt Brace Jovanovich, particularly Matt Milan, Lee Shenkman, Richard Lewis, and Roberta Astroff. Their suggestions helped to improve the manuscript. I alone, however, am responsible for the opinions expressed and for any errors that remain.

Finally, I wish to thank my students. Without them, the book would not have been possible, and the last eight years would not have been as exciting or rewarding.

I also owe a debt to other books. Those I have found most useful I have listed in the bibliography on page 135. From one of them, Strunk-White's *The Elements of Style*, I borrowed the device of using axioms in a short book, a format that has always appealed to me. In every other respect, however, the two books are different.

For me, writing has always been a way of learning. I started this book with a few key concepts and 63 specific axioms. As I wrote and learned, some concepts were recast or modified, and axioms were added or dropped, until only 50 remained. Looking back now, I realize how much I learned as I wrote this book, and I am more than ever conscious of all that still remains to be learned.

PREFACE viii

ACKNOWLEDGMENTS xi

FIGURES xviii

TABLES xx

1 PRINCIPLES 1

1–1 Technical writing is chiefly writing, and only secondly technical. 2
1–2 The basic function of technical writing is to inform. 2
1–3 The form and tone of technical writing depend on its purpose and audience. 3
1–4 Accuracy is the technical writer's first responsibility and the beginning of effective writing. 10
1–5 The essential qualities of technical writing are clarity, coherence, and conciseness—in that order. 10
1–6 Technical writing describes objects and explains processes, theories, and policies—usually in combination. 13

EXERCISES 17

2 WORDS 18

2–1 Use specific words in description for maximum clarity and power. 19
2–2 Know the meaning of each word used and consider its possible impact on the reader. 19
2–3 Use an adjective or an adverb only if it adds precision or dimension to the word it modifies. 22

CONTENTS

2–4 Do not use a long word when a short one will do the job. 23
2–5 Avoid clichés, euphemisms and jargon; the first are lifeless, the second deceptive, the third obscure. 24
2–6 Use a dictionary to ensure accuracy, a thesaurus to increase vocabulary. 27

EXERCISES 29

3 SENTENCES 31

3–1 A sentence is complete if it contains a subject and predicate and makes sense. 33
3–2 Verbs should agree with their subjects, pronouns with their antecedents. 43
3–3 Replace static verb phrases with strong verbs for more direct, forceful sentences. 49
3–4 Choose the active voice for clarity and force, the passive voice to shift emphasis or change the pace. 50
3–5 Avoid sentence structure that distorts or destroys meaning. 53
3–6 Vary the length and structure of sentences to avoid monotony and make them more effective. 55
3–7 Use parallel structure to stress the similarity between like ideas, objects, or processes. 58
3–8 Recast a sentence that seems weak or awkward; if it still falters, question its relevance. 59

EXERCISES 59

4 ORGANIZATION 00

4–1 To organize the material of a subject, first break it down into its component aspects. 63

4–2 To organize a report or paper, choose a suitable approach and make an outline that implements it. 64
4–3 The basic unit of organization is the paragraph. 66

EXERCISES 68

5 PUNCTUATION 71

5–1 The four main uses of punctuation are to end sentences and to introduce, separate, or enclose letters, words, phrases, or clauses. 72
5–2 The comma separates, encloses, and introduces; but its main use is to separate. 74
5–3 Do not confuse contractions with words that look alike or sound alike. 81
5–4 What follows the colon explains what precedes it. 83
5–5 The semicolon separates; do not confuse it with the colon, which introduces. 84
5–6 The hyphen separates letters and words; the dash separates words and phrases and indicates emphasis or abruptness. 85
5–7 Use parentheses to enclose the writer's words, brackets to enclose the editor's. 89
5–8 Place a comma or period inside quote marks; a colon or semicolon outside. 90

EXERCISES 91

6 FORMAT 93

6–1 Titles should be precise and descriptive, even long if necessary, but not cute or clever. 94
6–2 Formal reports are long, inclusive and elaborate; informal reports are less so. 95

6–3 The usual form of an informal report is a letter or memo; the memo is the more informal of the two. 97
6–4 Do not discuss theory or transcribe every detail of a lab notebook in a lab report. 109
6–5 An abstract explains what a report is about; a summary condenses what a report says. 110
6–6 Use tables to present related data, graphs to show trends, and drawings to clarify descriptions. 113
6–7 Spell out approximate numbers, numbers under 10, and numbers that begin a sentence; otherwise use numerals. 123
6–8 Place footnotes at the bottom of the page and a bibliography at the end of a report or paper. 124

EXERCISES 136

7 STYLE 138

7–1 Be selective: focus on the essential information, the significant detail. 139
7–2 Develop a lean, direct style; avoid inflated language and rambling sentences. 139
7–3 Write in the present tense whenever possible for simplicity and immediacy. 142
7–4 Use examples and comparisons to clarify descriptions and explanations. 146
7–5 Repeat words or phrases for clarity or emphasis, or to ease transitions; but avoid needless repetition. 147
7–6 Delete needless words and phrases, but avoid shortcuts that sacrifice meaning. 151
7–7 Choose clarity over style if they conflict. 153

EXERCISES 153

8 METHOD 155

8–1 Plan a report or paper thoroughly before starting to write it. 156
8–2 Gather the necessary data: basic library research. 158
8–3 Write a first draft. 177
8–4 Revise and rewrite—as often as necessary. 177

EXERCISES 178

APPENDIX 1 ANALYSIS OF A SAMPLE REPORT 179

APPENDIX 2 OVERVIEW OF GRAMMAR 184

APPENDIX 3 OVERVIEW OF SPELLING 190

APPENDIX 4 CONVERSION TABLES 194

INDEX 199

1–1 A Sample Proposal in Memo Form 4–5
1–2 Sample Work Procedure Instructions 8–9

4–1 Organizing Random Material 63

6–1 A Sample Letter Format 98
6–2 A Sample Letter of Inquiry 102
6–3 A Sample Complaint Letter 103
6–4 A Sample Letter of Adjustment 104
6–5 A Sample Data Sheet (Resumé) 107
6–6 An Abstract of Section 6, "Format," of *The Elements of Technical Writing* 111
6–7 A Summary of Section 6, "Format," of *The Elements of Technical Writing* 112
6–8 United States Yearly Inflation Rate, 1960–1978 115
6–9 Rainfall in Sonoma County, 1975–1978 116
6–10 The Annual Federal Budget, 1970–1977 117
6–11 Flow Chart of the Process of Writing 118
6–12 Schematic Diagram of V-F Converter 119
6–13 Organizational Chart of a Typical College 120
6–14 Exploded-view Drawing of the Stop Gear Replacement for HP Model G, H, and J532A 121
6–15 Map of the United States, Showing Census Divisions and Regions 122

7–1 Paring Down an Inflated Sentence 141
7–2 The Time Spectrum Covered by the Six English Tenses 145

8–1 A Typical Library Catalog Card 163
8–2 A Bibliographical Notesheet 175
8–3 A Sample Note Card 175
8–4 Formal Outlines: Traditional and Decimal Formats 176

A–1 A Sample Memo Report 181–82
A–2 The Relationship of the Four Types of Meaning Words and How Each One Functions 186
A–3 A Comparision of Fahrenheit and Celsius Temperature Scales 195

3–1 Coordinating Conjunctions and Conjunctive Adverbs 41
3–2 Pronouns 46
3–3 Subordinating Conjunctions 57

5–1 Uses of Punctuation Marks 72
5–2 Commonly Used Contractions 81
5–3 Contractions Commonly Confused with Similar Words 82
5–4 Common Prefixes: Their Use and Meaning 86

5–5 Common Suffixes: Their Use and Meaning 86

6–1 Format of Formal and Informal Reports Compared 95
6–2 Projected Requirements and Job Openings 114

7–1 The Six English Tenses and the Progressive Form of Each 143
7–2 Transitional Words and Phrases 150

8–1 What Managers Want to Know 157
8–2 The 10 Basic Subject Classes in the Dewey Decimal Classification System 159
8–3 The 10 Divisions of the 600 Class, "Technology and Applied Sciences," in the Dewey Decimal Classification System 159
8–4 The Subdivisions of the 620 Division, "Engineering and Allied Operations," in the Dewey Decimal Classification System 160
8–5 The 20 Basic Subject Classes in the Library of Congress Classification System 161
8–6 The 17 Divisions of the T Class, "Technology," in the Library of Congress Classification System 162

TABLES

8–7 The Subdivisions of the TA Division, "Engineering. Civil Engineering," in the Library of Congress Classification System 162

A–1 The Forms and Functions of Verbals 187
A–2 Functions of Phrases and Clauses in Sentences 189
A–3 Conversion Table: U.S. to Metric 196
A–4 Conversion Table: Metric to U.S. 197

The Elements of Technical Writing

The principles that evolve from any activity depend on the nature of the activity. But the nature of technical writing is elusive. What is it that distinguishes it from other writing? Precision? Formality? Thought processes? Language? Seriousness? Objectivity? All these and more have been cited; still the distinctive difference between technical writing and other writing remains obscure.

What does the word *technical* itself tell us? *Webster's Third New International Dictionary* defines it as "having special, usually practical, knowledge." The origins of the word reflect this. *Technical* is derived from the Greek *tekhnē* and *tektōn*, which mean "craft" and "carpenter." Its Latin roots, *texere* and *tegere*, mean "to weave" and "to cover" (as in roofing).

Carpentry, weaving, and roofing are applied skills that require special knowledge. Even today the mark of a technical field is special knowledge, which usually is applied to design, production, or research. This suggests that technical writing is writing that involves special knowledge and its application and that this distinction governs its essential qualities.

The following principles are based on this premise.

1-1 TECHNICAL WRITING IS CHIEFLY WRITING, AND ONLY SECONDLY TECHNICAL.

This principle cannot be stressed enough. The format and terminology of technical writing have been emphasized to the point of forgetting that the activity itself is writing.

Technical writing requires special knowledge, and often a special vocabulary. That is its distinctive characteristic—not objectivity, accuracy, precision, complexity, or seriousness, as some technical writers suggest. Objectivity, accuracy, and the rest are typical of technical writing, but they also may be found in any well-written essay.

A technical writer may be asked to write reports, papers, articles, brochures, manuals, press releases, presentations, speeches, even film scripts. Whatever the nature of the assignment, however, or the special knowledge involved, the basic task is *writing*.

To write is to select and arrange words. And whether those words describe Walden Pond or the design of a rocket engine, their choice and order remain the writer's essential problems.

1-2 THE BASIC FUNCTION OF TECHNICAL WRITING IS TO INFORM.

Writing that informs rests on *description* and *explanation* (also called *exposition*). The key to description is specific detail; the basis of explanation is logical analysis. Description and explanation usually complement each other, as in this passage describing and explaining double refraction in a gemstone:

> A gem is doubly refracting when the top and side faces of its crystals are unlike—as they are in all crystal systems except the cubic. Light entering the crystal from its top face may find the mineral's atoms packed closely together. Light entering the same crystal from a side face may find the mineral's atoms spaced farther apart. So the light passing from top to bottom is slowed down more than the light passing from one side to the other, creating double refraction.

Technical writing may also instruct or persuade, as in instruction manuals or sales literature. But instruction and persuasion are essentially applications of information—special ways of informing that rely on emphasis, tone, and authority for their effect.

1-3 THE FORM AND TONE OF TECHNICAL WRITING DEPEND ON ITS PURPOSE AND AUDIENCE.

Purpose

Every piece of technical writing is designed to accomplish something. It may be to inform, instruct, or persuade, or to obtain something—or a combination of these. To accomplish these purposes, various categories of technical writing have been devised. Their number is indefinite, but their range can be narrowed to ten inclusive types. The basic format for the first eight is the formal or informal report. **(See 6-2, p. 95, and 6-3, p. 97, for the elements of reports, and Appendix 1, p. 179, for an analysis of a report.)**

1. **PROPOSAL** A suggestion for action, usually involving change or performance. It may be to solve a problem, bid for a grant or contract, suggest a new building site, revise policy, initiate research, or terminate a project. Proposals usually involve money, either as a cost factor or as a fee.

The audience for a proposal may be within a company or outside it. Where the audience is outside the company, the proposal usually is more formal and—as in a bid for a contract—more sales oriented.

Some proposals are solicited, most are not. Solicited proposals may include a list of *specifications*—specific, precise requirements (standards, dimensions, tolerances, schedules) for all designs, materials, and operations involved in a project. Military specifications are explained in *Specifications, Types and Forms* and *Standardization Policies, Procedures and Instructions.* Both are available from the Naval Publications and Forms Center, 5801 Tabor Avenue, Philadelphia, Pennsylvania 19120. For non-military specifications, see the *Federal Standardization Handbook,* available from the Government Printing Office, Washington, D.C. 20401.

Figure 1-1 shows a sample proposal that also contains elements of a survey report. Technical terms

FIGURE 1-1. A SAMPLE PROPOSAL IN MEMO FORM

To: Harmony Valley #56 Homeowners Association

From: Roseanna Thompson
Architectural Control Chairperson
Unit #14

Date: January 7, 1979

Subject: HARMONY VALLEY #56
LANDSLIDE RECONSTRUCTION RECOMMENDATIONS

This report presents the results of a preliminary study of a landslide which occurred last winter in the Harmony Valley #56 Development at the end of Brooks Lane adjacent to Unit # 55. The purpose of this study was to examine the exposed features at the site, to review the geologic data, and to recommend possible remedial action.

SURVEY RESULTS

On October 2, 1978, I inspected the exposed surface features of the existing landslide. No subsurface exploration was authorized or performed. A soil report on this area, however, by Merrifield Associates, dated March 13, 1978, indicates that the bedrock is a highly faulted, sheared, and altered heterogeneous mixture of sandstone, shale, chert, greenstone, and schist commonly known as Franciscan melange. The melange includes large zones of firm competent rock intermixed with zones of greenstone and shale altered to the consistency of clay. The rock weathers, yielding a clay-rich soil that is moderately expansive; that is, it experiences volumetric changes with changes in moisture content. Expansive soils are generally weak when saturated and relatively hard when dry.

Numerous minor faults, such as are typical in melange, are present in this area; however, the geologic maps consulted do not indicate active faults at this site. The landslides in this area are predominantly in the soil above the rock. This landslide is covered with colluvium [slopewash]. Fans of colluvial and alluvial [clay] material are apparent at the lower end of the drainage swale [a low area of land] . No soil on the concrete is evident.

There is no structural damage as a result of this slide, only visual damage to the common area. Unit #55 adjacent to the landslide is probably in no danger even if further sliding occurs. On October 2, 1978, I placed 17 survey stakes in and around the landslide and took their elevations. On December 13, 1978, I again took their elevations; as indicated on the attached survey notes, there had been no significant movement of the earth even after several days of rain.

have been explained in brackets. The survey notes mentioned in the report have been omitted to conserve space.

CONCLUSIONS AND RECOMMENDATIONS

While there are a number of possible remedies for landslides, the most important is drainage. Unstable masses can be made stable by making and keeping them comparatively dry. Drainage is the logical means of control and, in the long run, the cheapest.

The problem here is one of intercepting both surface water and ground water before they can reach the mass that is subject to slide, and of draining the mass in order to stabilize it. I suggest the following:

1. Construct a concrete v-ditch along the top of the slide to divert surface water flowing down towards the mass.

2. Build a concrete retaining wall at the bottom of the slope and intrench a long perforated corrugated metal drain pipe in the backfill behind the retaining wall. This pipe will collect ground water and drain it out behind the wall to a solid pipe connected to the storm drain in the street.

3. If possible, excavate the slide material and replace it with a more stable, well-drained material. A more economical solution would be to weight the toe of the landslide with rip-rap [large rocks] after excavation of the loose slide material. But if any tendency to slide still exists, rocks or boulders of any kind are inefficient.

After reconstruction, protection of the disturbed slopes from concentrated runoff is necessary. This protection can be achieved by planting fast growing, deep-rooted ground cover to reduce sloughing and erosion. Seeding or sodding is almost as effective for stopping very minor slides.

This landslide may be in equilibrium; however, I recommend that no further landscape watering be done in this area. The largest contributing factor to this slide is the presence of water in the soil. Without drainage, it could seep through to the slip plane and hasten any further movement of the earth. Soil test borings should be taken at a number of typical places to locate the water table before construction design begins.

With the winter season approaching, construction should begin after the rains. Meanwhile, I will continue my survey to watch for any change in the slide during the wet season. Emergency slide prevention measures can be taken if any drastic movement occurs.

2. FEASIBILITY REPORT An examination of the pros and cons of a proposal. Feasibility reports (also called *studies*) rely on analysis and persuasion. They attempt to answer two questions: Can it be done? Should it be done? If the decision is that it can and should be done, the report also addresses the questions of why, when, where, and how. The most important factors usually are: availability and capability of facilities, personnel, and materials, and amount of time and expense required. In a feasibility report, present all relevant data, calculate the probability of success, and recommend an action or, if necessary, further study of the problem.

Proposals and the feasibility of implementing them often are combined in one report.

3. SURVEY REPORT A thorough study of any subject. Some subjects of surveys are: potential markets for products, community resources, labor policies, market penetration, and public opinion. A poll is one kind of survey; the study of a possible site for a new plant is another. Survey reports usually evaluate something. They require accuracy, specific details of what was done, and careful analysis of probabilities.

4. PROGRESS REPORT An account of what has been accomplished on a project over a specific period of time and what may be expected in the next period. It may include changes in procedures or adjustments in schedules. Some progress reports record personal or professional development, as in psychotherapy or personnel evaluation.

A *status report* (also called a *project report*) is similar; the difference lies in the time covered. Instead of accounting for a period of time, it describes the status of a project at a particular time, usually the present. Status reports also describe the condition of a department, company, profession, or industry. In these cases, personnel may be evaluated on the basis of their work.

The annual report to stockholders by corporations is a type of status report. So is the annual State of the Union message delivered to Congress by the President.

5. COMPLAINT REPORT A critical assessment of an

action, policy, procedure, or person. It usually recommends an alternative or adjustment. A *trouble report* is similar, but it is more descriptive and analytical than critical. So is an *investigative report,* and an *accident report.*

6. TRIP REPORT An account of a business or professional trip. It records specific (and significant) places, events, conversations, and people met. A trip report attempts to answer the questions where, when, what, why, and who. It may include recommendations.

7. CONFERENCE REPORT A summary of a business or professional meeting, stating the time, place, subject, and personnel of the meeting as well as any results. Conference and trip reports often are combined.

8. LABORATORY REPORT A record of procedures and results of lab tests. This should not be confused with the *lab notebook,* which is the *complete* working record of the tests and the source of the more selective lab report. A lab report describes the scope of a project, the equipment used, the procedures followed, the results of tests, and any conclusions or recommendations. A *research report* does the same, but it may describe research other than lab tests. **(See 6-4, p. 109, for the format and organization of a lab report.)**

9. TECHNICAL PAPER A research paper written for a professional journal or magazine. Technical papers usually describe a theory or new development. They resemble technical reports in most respects. The main difference is that the audience for the technical paper usually is larger and more diverse. *Technical articles* are similar but usually less theoretical and more concerned with practical applications.

10. INSTRUCTION MANUAL Directions for work procedures or policies, or for the use of technical equipment or appliances. Instruction relies on clear, specific, complete directions presented in sequential order. Descriptions of complicated step-by-step procedures should be accompanied by drawings.

Manuals may be written in a formal or informal style, depending on the audience, but they should inspire

FIGURE 1-2. SAMPLE WORK PROCEDURE INSTRUCTIONS

OBTAINING A GOOD SOLDER JOINT
USING A SOLDER IRON

This procedure is intended for the skilled solderer. The intent is to offer added technical soldering knowledge.

Soldering is the joining of metal parts with an alloy that has a lower melting temperature than the metal parts being joined. Solder is a fusible alloy normally consisting of tin and lead and used for the purpose of metallurgically joining together two or more metals.

Flux, a chemical substance used for cleaning and preparing the surface to be soldered with molten solder, removes the oxide film contamination from the metals to be joined.

Oxide film results from the combination of metal and oxygen when a clean metal surface is exposed to air.

The procedures for obtaining a good solder joint using a solder iron are:

1. Clean metal parts to be soldered. Flux is intended to remove oxide film, not dirt or grease. Dirt or grease must be removed with Freon or related solvents.

2. Clean soldering iron tip and "tin" all faces of tip with a coating of solder. This will reduce oxide film contamination from forming on the solder iron tip during idle time.

3. Heat parts with solder iron. Solder should be applied to heated parts until it melts and flows freely. Practice to develop a short soldering dwell time. Too much heat sometimes damages parts.

4. If separate soldering flux-liquid or paste is used, it should be applied with a swab or brush. With flux-core solder, separate fluxing is usually necessary, because flux is combined within the solder.

confidence. Where equipment or a procedure is dangerous, the reader's safety should be the first consideration.

Typical work procedures instructions (in this case, for skilled solderers) are shown in Figure 1-2.

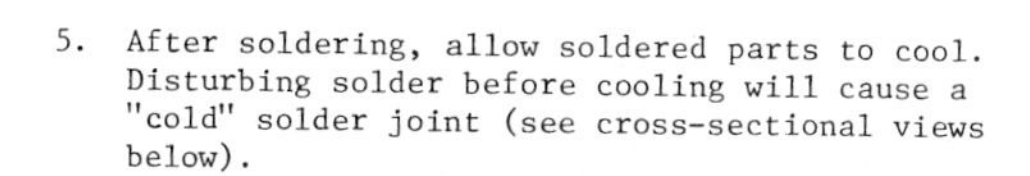

5. After soldering, allow soldered parts to cool. Disturbing solder before cooling will cause a "cold" solder joint (see cross-sectional views below).

6. A solder joint is good only when the angle is shallow and this angle is at all edges of the solder. Illustrated below are cross-sectional views of a good solder joint, and five poor joints resulting from commonly made mistakes:

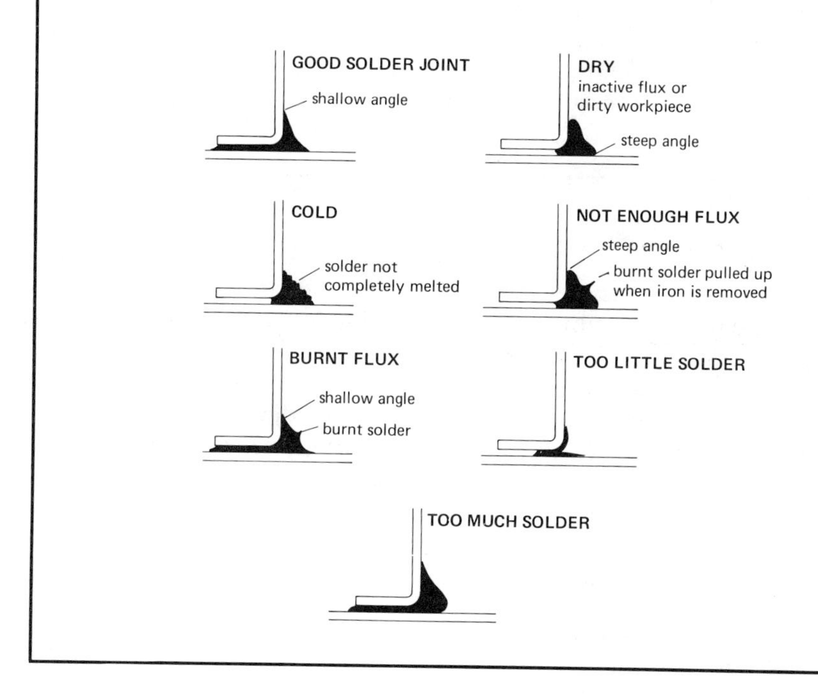

Audience

The specific audience of a piece of technical writing is the person or persons it is written for. You should know (1) who are they? (2) what do they know about the subject? (3) what do they want? and (4) what do they want it for?

Use the answers to questions three and four to determine what to cover and to emphasize.

Use the answers to questions one and two to set the tone of the writing and its technical level. For example, the tone of a report for a supervisor should differ from that for a subordinate. The technical level should vary in writing for a professional journal, a business manager, a machine operator, or the public.

In a report for an individual, adjust the tone and the technical level to that person. For an audience of several, broaden the tone. If the technical level of an audience varies widely, write the report on the more technical level and preface it with a less technical summary. **(See also 3-6, p. 55.)**

1-4 ACCURACY IS THE TECHNICAL WRITER'S FIRST RESPONSIBILITY AND THE BEGINNING OF EFFECTIVE WRITING.

Technical writing is accurate when its information is correct and complete, and when it fulfills its specific purpose and is directed to its specific audience. Anything less is at best ineffective, at worst disastrous.

The foundation of accuracy is solid research, careful transcription of notes, thorough checking of figures, and a careful reading of each draft for *what the words say*. **(See also 8-3, p. 177; 8-4, p. 177.)** There are many acceptable ways to write a sentence, but no acceptable alternative to accuracy.

1-5 THE ESSENTIAL QUALITIES OF TECHNICAL WRITING ARE CLARITY, COHERENCE, AND CONCISENESS—IN THAT ORDER.

Clarity

Writing is clear when the reader can grasp its meaning on the first reading. The writer, not the reader, bears this responsibility.

Clarity begins where obstruction ends. Some common obstructions are: wordiness, vague or ambiguous words, words that are misused or misplaced, euphemisms and jargon, and the passive voice and static verb phrases. (These are discussed separately below and in section 2.) Note the following passages:

Acoustical problems are realized due to the lack of proper insulation.	The lack of proper insulation causes acoustical problems.
Unfortunately, our competitors have adopted this procedure, but we have not.	Our competitors have adopted this procedure, but unfortunately we have not.
When the other sections ask for budget information, it is my expectation that the Budget Director relate to them regarding their informational needs in a method that is understandable as far as terminology is concerned.	When other department heads ask the Budget Director for information, I expect him to answer them in language they can understand.

In the first passage, the verb phrase "are realized" is misused and in the passive voice. In the second passage, "unfortunately" is misplaced. In the third passage, "sections" is vague, "it is my expectation" is a static verb phrase, "relate to them" is ambiguous, and the rest of the clause is *gobbledygook,* the jargon of bureaucrats.

Behind clear writing lie clear thinking and thorough planning. Before writing, decide what you want to say. Then say it simply and directly. Use relevant comparisons and develop your material logically.

Coherence

To cohere is to stick together. Coherent writing flows logically and smoothly from point to point, sentence to sentence, paragraph to paragraph. All the parts hang together to form a whole that makes sense. Writing that doesn't hang together or doesn't make sense is incoherent.

Note the difference between the incoherent and coherent versions of the same passage:

INCOHERENT

All calls appear to be handled simultaneously, but the two memory devices make it possible for the system to operate so quickly. The data on how to connect calls is contained in the program store, which exchanges information with the call store Central control processes the information of both stores. All the calls in progress are recorded in the call store, but the system executes only one instruction at a time. Stored program control uses these two memory devices.

COHERENT

Stored program control uses two memory devices, a program store and a call store. The program store contains the data required to instruct central control how to connect calls and the call store keeps track of all the calls in progress. The two memory stores exchange information and feed instructions to central control, which processes them. Although the system executes only one instruction at a time, it operates so quickly—every two-tenths of a second—it appears to handle all calls simultaneously.

The incoherent passage is choppy; related information is scattered; and sentences contrast either illogical or irrelevant elements. The sentences do not hang together as they do in the coherent passage, which is smoothly and logically developed.

Conciseness

The expression of much in few words is the essence of conciseness. To write concisely is to omit the wordiness that obscures clarity and weakens writing.

Most technical writing is too wordy by 10 to 60 percent. This is true not only of student work but of professional and published work as well. In a recent study, when 182 senior executives and department heads were asked to name the most persistent flaws in the technical writing they read, 90 percent of them complained of wordiness. The sentence below, for example, is 53 percent wordy. Imagine ten pages of such sentences.

The reason that quality control over the past several years has gone down is that management has not taken the trouble to train inspectors enough to do the job.	Quality control has suffered in recent years because management has not trained inspectors adequately.

Concise writing is brief and to the point—every word counts. The standards of conciseness are *relevance* and *necessity*. If an idea, word, or detail is irrelevant or unnecessary, drop it. For example, in writing a report about personnel policies of a company, do not include material about customer relations or sales techniques.

Conciseness begins with focusing on the relevant ideas, the necessary information, the essential words. But in the end you achieve it by editing as many drafts as necessary. Antoine de Saint-Exupery, a French writer who was a master at expressing much in few words, said that perfection is finally attained "not when there is no longer anything to add, but when there is no longer anything to take away." **(See also 7-6, p. 151.)**

1-6 TECHNICAL WRITING DESCRIBES OBJECTS AND EXPLAINS PROCESSES, THEORIES, AND POLICIES—USUALLY IN COMBINATION.

Special knowledge, the distinctive quality of technical writing, ranges from fashion design to aerodynamics. But whatever the technical knowledge may be, the writing itself describes or explains some combination of object, process, theory, or policy.

Guidelines for describing each of these follow.

Object

To describe a simple object, explain (1) what it is, (2) what it does, and (3) what it looks like—its size, shape, material, and finish.

Define any terms that may be unfamiliar to the reader. Use comparisons, contrasts, and examples to help the reader understand what the object does and looks like. Include drawings or photographs.

To describe a complicated object, begin with an overview of what it is and does and looks like. Then break it down into its major parts. Describe each part as above (name, function, appearance) and also explain its relationship to any part it affects. If necessary, break each part down into its important features and describe them separately. As in all description, include necessary definitions, and use comparisons, examples, and illustrations.

"The most difficult thing in writing," says Gore Vidal, the distinguished critic, playwright, and novelist, "is to describe something—whether it be an emotion or quite literally a table." Note the detail involved in the description of even a simple die:

> A die is a solid cube that is cast on a flat surface, usually in pairs, in games of chance. Carved in bone or molded in plastic, it ranges in size from 1.27 to 1.58 cm square, and usually is white, green, or red. Each face of the die is marked with from 1 to 6 spots incised 0.040 cm into the surface of the cube. The spots are arranged symmetrically on each face and placed so that those on opposite faces add up to 7.

Process

To explain a process, describe (1) what it is, (2) what materials and equipment it involves, (3) what happens, (4) how it happens, (5) why it happens, and (6) any results. If people are involved in a process, describe their role.

Emphasize what happens and how and why, except when describing materials and equipment. These are objects, so describe what they are, what they do, and what they look like.

To explain a complex process, divide it into its major steps and describe each one in sequence. If a process is cyclical (such as the nitrogen cycle, in which vital nitrogen compounds circulate from the soil to plants to people and back to the soil), choose a phase of it as a starting point.

As in the description of objects, include necessary definitions and use examples, comparisons, drawings, and photographs.

In the paragraph below, Charles F. Stevens explains how nerve impulses are transmitted from neuron to neuron in the brain. The *axon* here is the single long fiber of the neuron; the *dendrites* are the neuron's multiple short fibers.

> The functioning of the brain depends on the flow of information through elaborate circuits consisting of networks of neurons. Information is transferred from one cell to another at specialized points of contact: the synapses. A typical neuron may have anywhere from 1000 to 10,000 synapses and may receive information from something like 1000 other neurons. Although synapses are most often made between the axon of one cell and the dendrite of another, there are other kinds of synaptic junction: between axon and axon, between dendrite and dendrite, and between axon and cell body.

Theory

A theory is a general principle offered to explain systematic observations. It usually is arrived at through the inductive or deductive method. Induction begins with specific observations that may lead to a general conclusion. Deduction begins with a general conclusion that is supported by specific observations. **(See also 4-2, p. 64).**

The description of a theory rests on logical analysis and objectivity (to the degree that objectivity is possible). Describe the observations and explain the principle involved: its scope, possibilities, probabilities, limitations, probable objections, and cause-and-effect relationships.

Define terms when necessary and use examples, comparisons, and illustrations to aid the reader.

The following passage is taken from Albert Einstein's explanation of the formula $E=mc^2$. Note the comparison between a rich miser and material that does not give off energy externally.

> It is customary to express the equivalence of mass and energy (though somewhat inexactly) by the formula $E=mc^2$, in which c represents the velocity of light, about 186,000 miles per second; E is the energy that is contained in a stationary body; m is the mass. The energy that belongs to the mass m is equal to this mass, multiplied by the square of the enormous speed of light—which is to say, a vast amount of energy for every unit of mass.
>
> But if every gram of material contains this tremendous energy, why did it go so long unnoticed? The answer is simple enough: so long as none of the energy is given off externally, it cannot be observed. It is as though a man who is fabulously rich should never spend or give away a cent; no one could tell how rich he was.

Policy

The explanation of a policy should be complete, persuasive, and authoritative. Describe the purpose, scope, directives, exceptions, and any benefits or penalties of the policy.

Clarity and directness are essential in describing policies. So is an understanding of how people behave, especially how they respond to change.

The paragraph below is excerpted from the Santa Rosa Junior College policy on academic freedom. Note that it describes responsibilities as well as rights.

> The teacher, under the principles of academic freedom, has the right to discuss in his classroom all issues, however controversial, that he considers relevant to the nature of his course. This right carries with it the responsibility of considering controversial issues objectively. While the teacher has the right to present conclusions to which he believes the

evidence points, he has the responsibility of acknowledging the existence of—and showing respect for—opposing opinions.

EXERCISES

1. Write a proposal to the president of the school advocating the evaluation of instructors by students. Explain why it should be done and how it can be done. Suggest 10 criteria by which instructors can be evaluated. Decide whether this should be a pilot project or a full-scale program immediately.

2. Write a survey report describing the parking situation on campus or in the downtown area. Support the description with statistics. In the conclusion, discuss the probabilities for the immediate future.

3. Write instructions on how to parallel park an automobile. Be sure to include every operation as well as safety precautions.

4. Write instructions on how to change a flat tire. Be sure to include safety precautions.

5. Assume that a potential employer has asked you to write a paragraph explaining the nature, function, and essential qualities of technical writing. Write such a paragraph as you would if a possible job depended on it.

6. Write a description of a particular tool—for example, your own hammer or screwdriver. Include a diagram.

7. Assume that the next class meeting is a conference of technical writers from various companies gathered to hear a lecture on a specific aspect of technical writing. Prepare a conference report of the meeting. Include the time and place of the conference, the name of the topic and speaker, and the relevant points covered in the lecture.

The word is the basic element of the sentence, and therefore of writing itself. English contains more than one million words (three-quarters of which are technical), but only about 20,000 of them are commonly used. Of these, one-fifth are Anglo-Saxon, including the 10 most frequently used: and, be, have, it, of, the, to, will, you, I. Three-fifths are borrowed or modified from French, Latin, and Greek. And the rest are taken from more than 50 languages around the world, especially the Indo-European family.

We actually have three vocabularies: a reading, a speaking, and a writing vocabulary. Although the three overlap, our reading vocabulary is the largest. It contains many words that we recognize but do not ordinarily use in speaking or writing. Because we speak more than we write, our speaking vocabulary is the next largest. That leaves our writing vocabulary as the smallest—one reason why writers often must struggle to find the words they want. We cannot always pull out of our memory the words we recognize in reading or use in speaking.

The spoken and written languages also differ in other respects. We write more formally than we speak, probably because the written word bears the burden of meaning. In conversation, the words themselves carry less than 10 percent of our message. The rest is conveyed by our facial expressions, tone of voice, gestures, and posture. In writing, however, the words must carry more than 90 percent of the message. The rest is conveyed by punctuation and graphics.

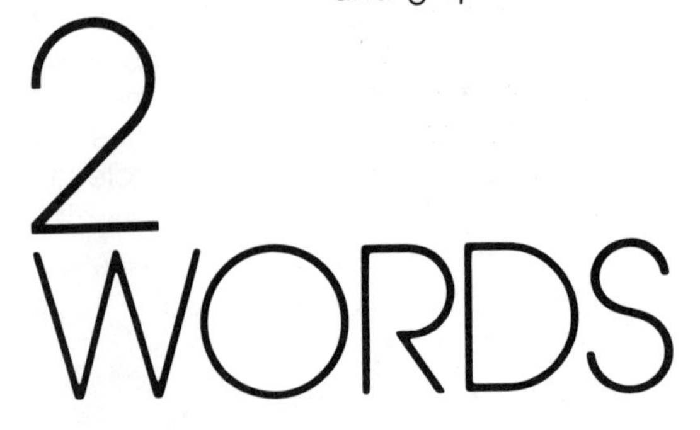

2-1 USE SPECIFIC WORDS IN DESCRIPTION FOR MAXIMUM CLARITY AND POWER.

In the study mentioned earlier, in which 182 executives and department heads were asked to name the most persistent flaws in the technical writing they read, every one of them cited vague language. The most effective antidote for vague language is specific words.

Not everything can be described specifically. Abstract concepts like freedom, justice, intelligence, and leadership defy specific description: they cannot be seen or touched, or even precisely defined. To describe such concepts beyond the abstract words for them, we must use examples—which remain arbitrary and debatable.

Abstract words perch on the top rung of what is often referrred to as the "ladder of abstraction." Below them sit general words like *machine, equipment, material, person:* concrete entities that can be seen and touched and precisely defined. But in description, general words still leave much to the reader's imagination. Is the machine a crane or a bulldozer, the equipment a flask or a test tube, the material steel or leather, the person a mechanic or an accountant? Generalities do not give the reader this information; the picture remains incomplete.

At the bottom of the ladder stand specific words like *lathe, microscope, stenographer, diode.* They create in the reader's mind distinct pictures, which sharpen and strengthen description. The reader does not have to supply any information; it is all there in the words themselves. To use specific words, then, is to give the reader the complete picture rather than just a portion of it.

2-2 KNOW THE MEANING OF EACH WORD USED AND CONSIDER ITS POSSIBLE IMPACT ON THE READER.

The writer's meaning is always subject to the reader's interpretation. This makes words potentially as misleading as they seem informative, since the same word may convey different meanings to different people. The word *run,* for example, has at least 145 separate meanings, some of which are difficult to distinguish. To reduce the possibility of misleading the reader, the writer must know the *nuances,* the shades of meaning, that words acquire.

Word meanings fall into two general categories: *denotative* and *connotative.*

The denotative meaning of a word is the dictionary definition. There may be several. For example, *liberal* is defined as: "of, relating to, or based on the liberal arts; generous, bountiful; not literal; not narrow in opinion or judgment; tolerant."

The connotative meaning of a word is the explicit dictionary definition plus an implied meaning acquired through usage. Here again, there may be more than one meaning. *Liberal,* for example, has acquired the connotation of "permissive," even "indulgent."

As attitudes and behavior change, gaps often appear between the denotative and connotative meanings of words. For example, the denotative meaning of *massage* is "remedial or hygienic treatment of the body by manipulation." This was the generally accepted understanding of *massage* before the rise of massage parlors that offer sexual services to their clients. Since then, the word itself has acquired a sexual connotation.

A connotative meaning that persists becomes a denotative meaning in a future edition of the dictionary. Thus today *radical* is defined as "fundamental," "thorough," and "extreme" as well as its original meaning, "going to the root of," which was derived from the Latin *radix,* "root." In medieval philosophy the *radical* sap was the vital moisture that was present in all plants and animals. But by the seventeenth century the word had acquired a political meaning: *radical* reform meant "fundamental" or "thorough" reform. By the nineteenth century, *radical* also had acquired the meaning of "extreme," having been used to describe the extreme left wing of the Liberal party in England. With the persistence of such usage, the connotations "thorough" and "extreme" became denotations.

Words also acquire positive or negative connotations according to the emotional responses they produce. Similar words may even have contrary connotations, one positive, the other negative. *Remedy* and *remedial,* for example, each produce a different emotional response. So do *inexpensive* and *cheap.*

Finally, there are those words that appear to have the same meaning but do not. Some of those that most often confuse writers are listed below.

1. **AFFECT/EFFECT** *Affect* is a verb that means "to influence"; *effect* is a noun and verb. The noun means the "result" or "consequence" of an action, as in "the *effect* of carbon monoxide poisoning." The verb means "to accomplish" or "to produce," as in "to *effect* repairs" or "to *effect* a solution." *Effect* is rarely used as a verb, and rightly so, since such expressions as "to effect repairs" and "to effect a solution" are at best awkward, at worst pretentious. "To *make* repairs" and "to *find* a solution" are clearer and more natural; *repairs* and *solves* are even clearer and more direct. (**See also 3-3, p. 49.**)

2. **AMOUNT/NUMBER** *Amount* refers to abstract quantity, as in "the *amount* of radiation allowable." *Number* refers to specific items that can be counted, as in "the *number* of colleges in California." Similarly, *less* refers to quantity and *fewer* refers to countable items, as in "the new computer breaks down *less* and makes *fewer* mistakes."

3. **ANXIOUS/EAGER** *Anxious* means "uneasy," as in *anxiety; eager* means "highly enthusiastic."

4. **BETWEEN/AMONG** *Between* refers to only two subjects; *among* to three or more.

5. **CAN/MAY** *Can* indicates ability to do something; *may* indicates permission to do it.

6. **CAPITOL/CAPITAL** The *capitol* is the building where the legislature meets. The *capital* is the political seat of government; a *capital* is the head of a pillar; *capital* is a form of wealth. When used as a modifier, *capital* indicates seriousness and importance, as in "*capital* letter," "*capital* punishment," and "*capital* crime."

7. **CENSOR/CENSURE** To *censor* is to suppress objectionable parts of something, such as a book or movie. To *censure* is to reprimand someone, such as a public official.

8. **CONTINUAL/CONTINUOUS** *Continual* means "steadily recurring" but with occasional interruption, as in

"continual disagreement between labor and management." *Continuous* means "continuing without interruption," as in "continuous efforts to release the hostages." *Constant,* a word often confused with these, means "continually recurring" but always in the same way and with similar results, as in "the *constant* repetition of mistakes."

9. CREDIBLE/CREDITABLE *Credible* means "believable"; *creditable* means "worthy of esteem or praise."

10. DISINTERESTED/UNINTERESTED *Disinterested* means "free from selfish motive or interest"; *uninterested* means "not interested."

11. IMPLY/INFER *Imply* means "to express indirectly; to suggest." *Infer* means "to derive as a conclusion from facts or premises; to guess." Writers imply; readers infer (and the result is often misunderstanding).

12. PRINCIPAL/PRINCIPLE Used as an adjective, *principal* means "most important." Used as a noun, it refers to a leading person (as in a play), the head of a school, or a capital sum of money that draws interest. A *principle* is "a general or fundamental law, doctrine or assumption."

In technical writing, the meaning is the message. To be sure the reader understands your message, use the right word in the right context. Anything less risks costly misunderstanding. As Mark Twain once said: "The difference between the right word and the almost right word is the difference between lightning and the lightning bug."

2-3 USE AN ADJECTIVE OR AN ADVERB ONLY IF IT ADDS PRECISION OR DIMENSION TO THE WORD IT MODIFIES.

Needless adjectives or adverbs weaken a sentence, weigh it down. This becomes clear if we examine the different functions and relative power of nouns and verbs and modifiers.

Nouns name things, verbs do things; they generate the power of a sentence. Adjectives modify nouns (and pronouns), and adverbs modify verbs (and adjectives and other adverbs). Modifiers limit the scope of the word modified. They indicate quantity, quality, time, or place, as in "*twin* gears," "*heavy* water," "arrive *tomorrow*," "move *forward*." They tell us something about the words they modify.

Modifiers are valuable, and sometimes necessary. For example, it may be necessary to indicate that a bridge is a *wooden* bridge or a *toll* bridge or a *suspension* bridge. To describe it as a *connecting* bridge, however, would be needless, a waste of space.

Every word takes up space and draws attention from every other word. If I write a single word on a page, that word commands the reader's full attention. Placing another word next to it divides that attention in two, diluting the power of the original word by half. Adding a third word dilutes the power of the other two proportionately. A fourth word dilutes them all still further, and so on.

In the same way, each modifier added to a sentence spreads the reader's attention more evenly throughout the sentence, diluting the natural power of the nouns and verbs. A necessary modifier presents no problem. Such words bring their own power to a sentence and add to its energy. But a needless modifier brings nothing to a sentence. It just sits there as dead weight, diluting the power of the nouns and verbs and weakening the sentence.

Used carefully, modifiers sharpen or enrich what you write. The test is whether they add to, or draw from, the words they modify. Avoid such vague modifiers as *very, undue, real, duly, nice, quite, rather, considerable, suitably, substantial, meaningful,* and *significantly*. They don't add enough to a sentence to justify the space they occupy. Look for stronger modifiers—or stronger nouns and verbs.

2-4 DO NOT USE A LONG WORD WHEN A SHORT ONE WILL DO THE JOB.

Three- or four-syllable words are often obscure, sometimes pretentious. One- or two-syllable words are usually more forceful than long words and easier to understand. The same holds true for plain instead of academic words, regardless of length.

Most long or academic words come from Latin or

Greek. Anglo-Saxon words are usually short and direct. Some commonly used long or academic words and their short or plain alternatives are listed below:

abrogate	cancel	ingest	swallow
ameliorate	improve	prognosticate	predict
cognizant	aware	propensity	bent
commence	start	refractory	stubborn
conflagration	fire	remuneration	pay
elucidate	explain	terminate	end
fabricate	make	unequivocal	clear
homologous	alike	utilization	use

Sometimes a long or academic word is necessary to convey a shade of meaning. For example, in the phrase "a facile solution," *facile* implies that the solution is shallow and reflects little thought, a meaning that the plainer word *easy* would not convey. Sometimes only a technical word, such as *photosynthesis,* will do the job. But look for the short, plain word first, the one the reader most likely will understand. Remember what the Eaglet said to the Dodo in *Alice in Wonderland:* "Speak English! I don't know the meaning of half those long words, and what's more, I don't believe you do either!"

2-5 AVOID CLICHÉS, EUPHEMISMS, AND JARGON; THE FIRST ARE LIFELESS, THE SECOND DECEPTIVE, THE THIRD OBSCURE.

Clichés

Clichés are commonplace expressions that have become trite or hackneyed by overuse. They may seem clever when first encountered, but they have been so worn down by repeated use they are flat and insipid. Poet Donald Hall calls them "little cinder blocks of crushed and reprocessed experience." Some common clichés:

explore every avenue
face the facts
for all intents and purposes
heart of the matter
height of absurdity
in the final analysis
it stands to reason

large body of information
last but not least
lay down the law
needless to say
needs no introduction
path of least resistance
pure and simple
rank and file
step in the right direction
too numerous to mention
voice an opinion
warrants further investigation
weigh in the balance

Each of these expressions can be either rewritten, shortened, or dropped. For example, "weigh in the balance" can be rewritten as "consider"; "last but not least" can be shortened to "last"; "it stands to reason" can be dropped with no loss of meaning.

Euphemisms

Euphemisms are words or expressions that make something appear better than it is or make it seem what it is not. The word comes from the Greek *euphanai,* "to speak well of." Euphemisms soften the edges of living and dying. Some examples, and the realities they disguise:

correctional facility	prison
demolition engineer	wrecker
dentures	false teeth
electronic surveillance	wiretap
facility event	nuclear accident
maintenance engineer	janitor
marketing representative	salesman
memorial park	cemetery
pacification	destruction
pass away	die
planned parenthood	contraception
previously owned	used
public relations counsel	publicist
realtor	real estate agent
restroom	toilet

senior citizen	old person
substandard housing	slum
terminal	fatal

In 1946 George Orwell, a British writer, wrote that euphemisms named things "without calling up the mental pictures of them." In other words, what you didn't see wouldn't hurt you. Such words as *facility event* and *pacification* are designed to gloss over the fear and horror associated with the events they refer to. Today, euphemisms often *create* an image—usually a safe or inflated one. That is, what you see is not what is. Words like *maintenance engineer* and *realtor* are concocted to raise the status of a particular occupation and of the people in it. Deception by omission or deception by substitution both violate the technical writer's first responsibility: accuracy.

Jargon

Jargon is unnecessarily obscure terminology peculiar to a profession or field. It should not be confused with indispensable technical terms, or even with shoptalk, which lacks the bloat of jargon. The term is derived from the French *gargon,* which means "any noise made in the throat."

Bureaucratic jargon is called *gobbledygook,* a term coined by a congressman who defined it as writing that is "long, pompous, vague, involved, usually with Latinized words."

The law, sociology, education, economics, the military, and government are the principal spawning grounds of jargon. Below is a sample of educational jargon taken from a high school principal's letter to parents:

> Our school's cross-graded, multi-ethnic, individualized learning program is designed to enhance the concept of an open-ended learning program with emphasis on a continuum of multi-ethnic, academically enriched learning using the identified intellectually gifted child as the agent or director of his own learning.

Writers often disagree on where to draw the line between necessarily technical language and jargon. In such cases the conclusive test is: It is jargon if it is unintelligible or if it can be restated simply and more concisely without loss of meaning.

Shoptalk like *hardware* (the mechanical equipment in the computer industry) and *galley* (a shallow metal tray used in type composition) cannot be simplified without adding several words. The principal's letter, however, is jargon. Even after several readings, its meaning can only be guessed. One attempt to translate it appears below:

> Our program for individual learning allows a gifted child of any ethnic background to help direct his or her education. An extension of the idea of open-ended learning, the program emphasizes academic development within the context of the culture of different ethnic groups.

"The great enemy of language," Orwell wrote, "is insincerity." Jonathan Swift, an eighteenth-century English writer, called it "affectation," a close relative. Behind the smoke of long words, one can generally find clichés, euphemisms, and jargon. For example, long or academic words, which imply learning (but do not necessarily reflect it), subtly cloak the writer with authority. To use clichés is to pass off as original someone else's cleverness, now stale. Euphemisms are seductively insincere, and jargon displays the writer's membership in a select group. Of course, there may be smoke without fire; but such smoke is no less suffocating.

2-6 USE A DICTIONARY TO ENSURE ACCURACY, A THESAURUS TO INCREASE VOCABULARY.

The dictionary defines, the thesaurus refines; the two work side by side. Consult the dictionary when in doubt about the meaning or root of a word; or how to use, spell, or pronounce it; or how to divide it into syllables. **(See also 5-6, p. 85, and Appendix 3, p. 190.)** For synonyms and antonyms, turn to the thesaurus.

Dictionaries

They differ in size, philosophy, and purpose. Unabridged dictionaries contain from 250,000 to 600,000 words; standard "desk" editions, from 150,000 to 180,000 words; and paperback editions, from 40,000 to 60,000 words. Desk and paperback dictionaries usually are more current than unabridged volumes because short dictionaries are cheaper to revise and reprint than long ones.

Lexicographers, people who compile dictionaries, disagree on what a dictionary should do. One school believes it should guard the language, prescribe standards of proper usage. *Webster's New International Dictionary of the English Language,* second edition, unabridged (1934), reflects this philosophy. Another school believes the guardians of the language are the people who use it, that common usage is the standard. *Webster's Third New International Dictionary of the English Language,* unabridged (1961), reflects this philosophy (and so does this book).

Most desk and paperback editions are patterned after one of these dictionaries or after the *Random House Dictionary of the English Language* (1966). One desk edition that is not is *Webster's New World Dictionary of the American Language,* second college edition, published by Collins-World (1978). It is strong on etymology (origins of words) and on American usage. And, like all good desk dictionaries, it explains the subtle distinctions among synonyms like *sly, cunning, crafty, tricky, foxy,* and *wily.* Among the paperback editions, the *Merriam-Webster Dictionary* (1974) carries the authority of its source, *Webster's Third.* Together, the *Webster's New World* and *Merriam-Webster* dictionaries define subtle distinctions and contemporary usage—a strong combination. The paperback edition offers additional guidance: any word not in it is likely to be beyond the understanding of many readers.

Two special dictionaries deserve mention. For the definition of a technical word or phrase, consult the *Dictionary of Scientific and Technical Terms* (1976). Covering 102 technical fields ranging from acoustics to zoology, it contains almost 100,000 entries, many of them illustrated. For the derivation of a word, see the great etymological dictionary of the English language, the *Oxford English Dictionary* (1961). The *OED* traces the roots and history of 500,000 words known to be used since 1150. It describes changes in meaning, spelling, pronunciation, and usage, and illustrates each meaning in the context of a sentence.

Thesaurus

The word *thesaurus* is derived from the Greek *thēsauros,* which means "treasury" or "storehouse." The original thesaurus of words was compiled by Peter Mark Roget, an English physician and scientist, and published in 1852. Roget's *Thesaurus* contained 15,000 entries classified according to key ideas and concepts. Since then, Roget's *Thesaurus* has been expanded, published in both British and American editions, and arranged in alphabetical order. Several American editions using Roget's original classifications are still available, but the editions arranged in dictionary form are easier and more convenient to use. The most popular of these is the *New American Roget's College Thesaurus* (1978), published in paperback by Signet.

Among technical writers—both student and professional—an inadequate *general* vocabulary is either the first or second most common handicap. To reduce that handicap, the thesaurus is the single best tool. But do not use it as a dictionary; use it *with* a dictionary. When you select a word from the thesaurus, check its definition to be sure it means exactly what you want to say.

EXERCISES

1. Describe a chess board and chess pieces in enough detail for a chess set to be designed from the description.
2. Describe how the classroom in which your class meets differs from another classroom (for example, a

lab) in which you meet. Be sure to include significant details.

3. Look up the word nice and any other word in the Oxford English Dictionary. Prepare a brief report on each, including its source, its earliest recorded use, and a record of its meaning and usage.
4. Rewrite the following passage, without the fancy words, jargon, and needless modifiers:

> It is imperative that we eliminate all unessential verbiage from our writing patterns, inasmuch as such superfluousness enhances the distinct probability of our not being comprehended by the ordinary reader.

5. Write a paragraph describing the uses of and the differences between a dictionary and a thesaurus.

6. List five technical or scientific words and report on their origin. (For example, telephone combines the Greek tele, meaning "far off," with phone, a variation of the Greek phono, meaning "sound" or "voice".)

3 SENTENCES

The sentence is the backbone of the language, the basic measure of writing. In it, specific words are combined in a specific order to convey a specific meaning. Word order, or sentence patterns, is the subject of syntax, the branch of grammar concerned with the relationships between words. (See Appendix 2, p. 184.)

Where syntax is concerned, perspective is all. Prescriptive grammarians base their authority on rules, with their insistence on form; descriptive grammarians base their authority on usage, with its reliance on meaning. English itself has evolved according to usage. From a synthetic language in which meaning depended largely on inflections—internal changes in the words themselves—English has grown into an analytic language, where meaning depends largely on phrases and word order. For example, in the sentences "Have you anything left?" and "Have you left anything?" the only difference is in the position of the word left. Yet the sentences pose different questions.

Old English (used from 450–1100) was an inflected language. The word *the*, for instance, had 17 distinct forms to indicate person, number, gender, and so on. Each form had to agree with the form of the noun it modified—which itself had to agree with the form of the verb it combined with. Such inflections were gradually leveled during the Middle English period (1100–1500) and eventually eliminated in Modern English (1500–present). Today only 5 broadly used inflections survive, all endings. These are: *-s* to indicate plurals and the third person singular of verbs; *-ing* and *-ed*, the present and past participle forms of most verbs; and *-er* and *-est*, the comparative and superlative forms of adjectives (*great*, *greater*, *greatest*).

Speakers of American English continue to simplify the language. For example, the use of *whom*, the objective form of *who*, steadily declines, despite the scolding of prescriptive grammarians. More changes will surely follow, for language reflects the life of the people who use it, and in life change is as constant as death is certain.

3-1 A SENTENCE IS COMPLETE IF IT CONTAINS A SUBJECT AND PREDICATE AND MAKES SENSE.

The sentence has been diagrammed, transformed, measured, graphed, and embedded—yet few writers understand it. Here it will be described in terms of (1) its basic *framework,* (2) its *elements* and how they combine and *function,* and (3) the sentence *patterns* in which these elements interact. All three affect meaning, clarity, and style; the third also affects emphasis. **(See also 3-6, p. 55, and Appendix 2, p. 184.)**

Inevitably, the description of sentences requires grammatical terms. Here these have been reduced to the necessary minimum and defined by function. The emphasis is on those constructions writers normally use.

Grammar becomes easier to understand if all words are divided into two categories: *meaning* words and *structure* words. The name of each category is itself definitive.

Meaning words are those words that convey the basic meaning of a sentence. There are only four: nouns, verbs, adjectives, and adverbs. Adjectives, which modify nouns, and adverbs, which modify verbs (or adjectives or other adverbs), are often simply called *modifiers.*

Structure words include all the others—pronouns, conjunctions, prepositions, and so on. They help hold the sentence together and, in doing so, add to its meaning and refine its style. *In this* sentence, *for* example, *all the* words *in* italics are structure words; *those in* roman type are meaning words. These eight structure words perform several vital functions in the sentence. Each *in* places the subject of the sentence, *words,* in a context: the words are *in* a sentence, *in* italics, *in* roman type. The largest context is limited to the sentence at hand by *this. For* connects the noun *example* to the noun *sentence,* indicating that the example is the sentence itself. *All* and *the* lead the reader to the noun *words,* providing a sort of continuity. Either one (but not both) could be dropped without breaking that continuity. Finally, *those,* a pronoun, stands in for the noun phrase *all the words,* which would sound monotonous if repeated. **(See also Appendix 2, p. 184.)**

The framework, elements, and patterns of sentences are examined below.

Sentence Framework

The basic framework of a sentence consists of a *subject* and *predicate* that stand alone and make sense.

The subject is what the verb makes a statement about. It may be a word, phrase, or clause. In the example below, the subject (underlined) is a noun phrase:

A number of writers attended the conference in Los Angeles.

The predicate consists of the verb and everything that follows it to complete the meaning of the sentence. In the example below, the predicate is underlined.

A number of writers attended the conference in Los Angeles.

What follows the verb may be: an *object,* a *predicate noun* or *adjective,* or a *modifier.*

The *object* of the verb is the person or thing that receives the action of the verb; it answers the question *what* or *who.* In the example above, the object is *the conference;* it answers the question of *what* the writers attended.

A *predicate noun* names the subject of the sentence again, but in different terms:

Dreyer is a writer.

A *predicate adjective* describes the subject of the sentence:

The flight to Los Angeles was rough.

A *modifier* may appear anywhere in a sentence. In the sentence below, it appears (underlined) after the verb:

The speaker lectured without notes.

Note that the modifier here is a phrase, one of the three basic elements of a sentence discussed below.

Sentence Elements

The three basic elements of a sentence are: *words, phrases,* and *clauses,* the first combining to produce the last two. The important distinction between phrases and clauses is that clauses contain a subject and predicate, and phrases do not.

Words, the basic unit of writing, function as nouns, verbs, or modifiers. Phrases and clauses, acting as single units, do the same job, with one exception—a clause, because it also contains a subject, cannot function as a verb.

Phrases take four basic forms, depending upon the key word in the phrase: prepositional, verbal, noun, or verb.

A prepositional phrase consists of a preposition (*for, to, of, in, from, after, with,* and so on) with a noun or pronoun as its object. Prepositional phrases usually function as modifiers. In the quote below, taken from the seventeenth-century French writer La Rochefoucauld, the phrase *at their true worth* modifies the noun *things:* it describes how things should be estimated.

"The greatest of all gifts is the power to estimate things at their true worth."

The term *verbal* describes the *infinitive, present participle,* and *past participle* forms of a verb (*to learn, learning, learned*) when they do not function as verbs. A verbal phrase may function as a noun, but usually functions as a modifier. In the Rochefoucauld quote above, *to estimate things* is a verbal phrase (infinitive) that modifies the noun *power:* it describes what the power is. Note that the prepositional phrase *at their true worth* is part of the complete verbal phrase *to estimate things at their true worth.* The prepositional phrase is said to be *embedded* in the larger phrase. This is common in sentences. Two, three, or more phrases may be embedded in one another or in a larger structure, a clause.

A noun phrase, a group of words based on a noun, always functions as a noun and as the subject of the sentence. In the example below, the noun phrase is underlined. The *headword,* or key noun, of the phrase is *automobiles,* for the essential message of the sentence is that new *automobiles* will be unveiled in the fall.

> The new model automobiles Detroit has been promoting will be unveiled in the fall.

A verb phrase consists of a helping verb(s) and main verb and always functions as a verb.[1] For example: *may have learned.*

Unlike a phrase, a *clause* contains a subject and a predicate. A clause able to stand alone and make sense is a complete sentence. It is also called an *independent* clause, in contrast to a *dependent* (or subordinate) clause. A dependent clause cannot stand alone and make sense; it depends on an independent clause to complete its meaning. The example below contains a dependent and independent clause. The dependent clause is underlined.

> Flight 907 arrived late because it departed New York late.

The independent clause carries the main message of the sentence: Flight 907 arrived late. Standing alone, the

[1] Do not confuse a verb phrase with a static verb phrase, which is discussed in 3-3. A static verb phrase results when a strong verb is converted to its noun form and attached to a weak verb. For example, *decide* becomes *decision* and is attached to the verb *make* to form the phrase *make a decision*—a needless expansion that weakens the sentence.

clause would still make sense. The dependent clause, however, would not make sense by itself. It depends on the independent, or *main,* clause for its context; without that, it means nothing. In fact, it modifies the verb of the independent clause: it explains why flight 907 *arrived* late. This dependent clause, then, is also an example of a clause that functions as a modifier, as discussed on page 188.

Sentence Patterns

Sentence patterns in English, like the words themselves, have been established by usage. Most common is the subject-predicate pattern (S-P). Three-quarters of all sentences follow this order. Why? Because the subject of a sentence is what the sentence is about, and users of English customarily place it in the attention-getting position: up front (and in the main clause).

Of the remaining quarter, most follow an order of modifier-subject-predicate (M-S-P), where the modifier is a word, phrase, or clause. Less than 1 percent of English sentences follow an inverted order, in which the predicate precedes the subject (P-S). An example of each pattern follows, with the subject underlined:

This sentence illustrates the S-P pattern.

In this sentence, the order is M-S-P.

Less common is this P-S sentence pattern.

In the first example, *the S-P pattern* is the object of the verb *illustrates.* In the second, *M-S-P* is the complement of the verb *is;* it renames the subject *(the order).* And in the third, *less common* is the complement of the verb *is;* it describes the subject *(this P-S sentence pattern).*

Correct Sentences

The most reliable measure of the correctness of a sentence is whether it makes sense. If it doesn't, the problem lies either in a slight shift of meaning caused by an unwitting distortion or omission of information or in the structure of the sentence itself.

The main source of distortion in sentences is inaccurate cause-and-effect relationships. **(Causal relationships are further discussed in 4–2, p. 64.)** For example, examine the following sentence:

> The air conditioner was turned on because the windows were closed.

Is the causal relationship here accurate? Would it not be more accurate to say that the air conditioner was turned on because the room was hot? It is, of course, possible that the room was hot because the windows were closed, but if that were so, the logical action would have been to open them, not turn on the air conditioner. In short, the heat, not the closed windows, was the basic cause for turning on the air conditioner. To focus on the heat itself is also to cover the possibility that the room would have been hot even with the windows open; in fact, it implies it. Therefore the only accurate causal relationship between the closed windows and the turning on of the air conditioner would require a reversal of the cause-and-effect here, as in the following sentence:

> The windows were closed because the air conditioner was on.

This is accurate because open windows reduce the capacity of an air conditioner to cool a room.

The omission of information can be traced to the writer's unconscious condensation of a sentence. For example, examine the following sentence:

> There is not enough equipment here to be a science room.

It contains a problem, but one so vague that many writers fail to identify it. The question is, What do the *words* say? A careful reading reveals they say that the equipment is a room. That is impossible of course.

Equipment can be *in* a room, but it cannot *be* a room. No doubt the writer also knows that and meant to say "There isn't enough equipment here *for the room* to be *used as* a science room" but, in condensing the sentence, omitted the italicized words.

The full version of the sentence makes sense, but it is awkward and repetitive. A few word changes (underlined) produce a smoother sentence:

> There isn't enough equipment here to make the room suitable for a science class.

The main structural problems involve *fragment* and *run-on* sentences. A fragment sentence is an incomplete sentence; it lacks a subject or a verb. A run-on sentence consists of two complete sentences joined together as one. An example of each follows:

> Examining a patient.

> The maximum allowable exposure is five rems per year even that may be unsafe.

Sometimes a sentence seems to make sense, but doesn't when you read what the words actually say. This problem usually appears in M-S-P sentence patterns where the modifier is a verbal phrase that does not clearly or logically refer to the subject of the sentence. Such a phrase is usually called a *dangling modifier*. For example:

> Looking south, the property extends 400 meters to the bank of the Russian River.

The verbal phrase *Looking south* does not logically refer to the subject of the sentence, which is *the property*. Property, of course, cannot look in any direction; only a person can do that. The person who is doing the looking (that is, the writer) must be inserted into the sentence:

> Looking south, I saw that the property extends 400 meters to the bank of the Russian River.

The verbal phrase now logically refers to the subject of the sentence, which is no longer *the property* but *I.*

Fragment sentences usually are just phrases or dependent clauses. Below are three examples of each and, in parentheses, the elements that would make them complete sentences:

> (We were scheduled) To return to Denver on Sunday.
>
> Instead of raising prices. (the company cut costs.)
>
> (The west end of the project is) Bordered by hardwood trees.
>
> (The crew had coffee) While their plane was refueled.
>
> (The blackout ended) As soon as the generator was repaired.
>
> Since we must import oil. (we should cut fuel consumption.)

An apparent fragment, however, may make sense in the context of the preceding sentences. *Like this one. And this one. Or even this one.* These last three sentences are fragment sentences, but the first sentence in this paragraph provides the context that gives them meaning and in a sense completes them. Fragment sentences like these are usually short and created deliberately for their dramatic effect.

Most run-on sentences fall into one of four types: (1) those that simply run together; (2) those joined *only* by a comma; (3) those connected *only* by a *coordinating conjunction,* such as *and, or, but,* and so on; and (4) those joined by *conjunctive adverbs,* which are conjunctions that not only connect the two sentences but modify the second one—words like *however, also, therefore,* and so on. A list of coordinating conjunctions and conjunctive adverbs appears in Table 3-1. **(See Table 3-3 on page 57 for a list of subordinating conjunctions—conjunctions used to connect dependent to independent clauses.)**

TABLE 3-1. COORDINATING CONJUNCTIONS AND CONJUNCTIVE ADVERBS

COORDINATING CONJUNCTIONS	CONJUNCTIVE ADVERBS	
(Join words, phrases, or independent clauses)	(Join only independent clauses)	
and	accordingly	later
but	afterward(s)	moreover
yet	also	nevertheless
or	besides	otherwise
nor	consequently	still
for*	earlier	then
so*	furthermore	therefore
	hence	thus
	however	

*For and so connect only independent clauses

An example of the four types of run-on sentence follows, with suggestions on how to separate each one:

(1) Photosynthesis requires light it can come from the sun or from artificial sources.

Two sentences here—*Photosynthesis requires light* and *it can come from the sun or from artificial sources*—have been run together. To separate them, choose one of three alternatives: (a) place a comma and the coordinating conjunction *and* after *light;* (b) place a semicolon after *light;* (c) place a period after *light* and capitalize the *i* in *it.*

The choice is largely a matter of preference and style; all are structurally sound. It is worth noting, however, that only the semicolon emphasizes the close relationship between the two sentences.

(2) Fats and carbohydrates contain carbon, hydrogen, and oxygen, proteins contain these elements plus nitrogen.

Here the two sentences—*Fats and carbohydrates*

contain carbon, hydrogen, and oxygen and *proteins contain these elements plus nitrogen*—are separated only by a comma, which is not enough. The alternatives here are almost identical to those in the first example: (a) place the coordinating conjunction *and* after *oxygen*; (b) place a semicolon after *oxygen*; (c) place a period after *oxygen* and capitalize the *p* of *proteins*. Again, the choice is largely one of preference and style.

(3) A lobster quickly dies in salt water but salmon can live in either fresh or salt water.

The two sentences here—*A lobster quickly dies in fresh water* and *salmon can live in either fresh or salt water*—are separated only by the coordinating conjunction *but,* which is not enough. Again, the alternatives are almost identical: (a) place a comma after *water;* (b) place a semicolon after *water;* (c) place a period after *water* and capitalize the *b* in *but*. Here, too, preference and style are decisive.

(4) Both red and blue light promote photosynthesis however red light is more effective.

Here the two sentences—*Both red and blue light promote photosynthesis* and *red light is more effective*—are separated only by the conjunctive adverb *however,* which is not enough. Where a conjunctive adverb appears between sentences, only a semicolon or period before the adverb will adequately separate the sentences. And since the adverb also modifies the second sentence, a comma is needed between the adverb and the second sentence. In the example, place the semicolon or period after *photosynthesis* and the comma after *however*. If a period is used, the *h* in *however* should, of course, be capitalized.

A final note: writers who do not easily recognize run-on sentences may find it helpful to read their work aloud, for the voice naturally drops where a sentence ends.

3-2 VERBS SHOULD AGREE WITH THEIR SUBJECTS, PRONOUNS WITH THEIR ANTECEDENTS.

The question of agreement commonly concerns *number* (singular or plural) in nouns and verbs—and then only in the present tense of the verb.[1] It involves, in most instances, one of the five remaining common inflections in English: the final *-s.*

The plural of most nouns is formed by adding an *-s* to the singular: *storm + s=storms.*[2] The verb also has only two forms in the present tense: the infinitive *(strike)* and the infinitive plus an *-s: strikes.* (Some verbs take *-es: goes.*)

Use the *-s* (or *-es*) form with singular nouns and singular third person pronouns (*he, she, it*): *storm strikes, he goes.* Otherwise use the infinitive form: *storms strike, I/you/they go.*

We usually solve such questions of agreement instinctively, guided by the natural sense of meaning we acquire through a lifetime of speaking English. But some combinations create doubt, even confusion, because they contain more than one subject or involve words in which form conflicts with meaning or usage. This category includes: (1) nouns such as *family,* which have a singular form but plural meaning; (2) nouns such as *politics,* which have a plural form but singular meaning; (3) the pronouns *none* and *any;* (4) long noun phrases containing singular and plural nouns; and (5) *compound subjects,* such as *assault and battery.*

[2] The verb *to be* is one exception; it has singular and plural forms in the past tense: *was, were.* Not discussed here, because it presents few problems in agreement between subject and verb, is *case,* which indicates when a noun or pronoun functions as a subject, object, or possessive. Also not discussed—and for the same reason—is *person:* the identification of the speaker (*I*), the spoken to (*you*), and the spoken about (*he/she/it*).

[3] There are fewer than 50 exceptions. Some do not inflect at all to form the plural: *deer, sheep, salmon, trout.* Others inflect internally: *men, women, feet, teeth, mice, geese.* A few form the plural by adding *-en: children, oxen, brethren* (which also inflects internally).

1. Nouns with a singular form and plural meaning, called *collective nouns,* represent a group of individuals who have something in common. Some often-used collective nouns are: *family, team, audience, class, crowd, committee, jury, faculty, staff, army, group.*

When the meaning of a collective noun focuses on the group as a single unit—which is most of the time—use a singular verb:

The team is in first place.

But use a plural verb when the meaning of a collective noun focuses on the individuals in the group:

Her family are Democrats, and his (family) are Republicans.

Also use a plural verb when such collective nouns as *number, group,* and *crowd* combine with a prepositional phrase to focus on the individuals in the group:

A number of people were promoted.

Data, another collective noun, can take either a singular or plural verb, but usage of the singular is increasing:

The data was gathered by the research department.

2. Nouns with a plural form and singular meaning function like collective nouns. Such nouns as *politics, economics,* and *physics* describe unique activities; and measurements of weight, time, money, or distance represent the single sum of individual units.

Here, too, meaning is decisive—use a singular verb with such nouns:

Fifty million dollars is a large investment.

Sixty miles is a long commute.

Physics is a required course for engineering students.

3. The pronoun *none* is equivalent to *no one* and

probably an oral contraction of it. Since *no one* is singular, prescriptive grammarians insist *none* should be singular, too, and should take a singular verb. But in usage it is either singular or plural, depending on its meaning, and the plural use is more common:

None of the original founders are left.

The word none is sometimes singular, more often plural.

The same applies to the pronoun *any,* which is probably an oral contraction of *anyone:*

Any of them know the way.

Where policy is involved, any is preferable to none.

4. Long noun phrases often contain singular and plural nouns, such as the phrase in the sentence

Any policy without emergency provisions in its restrictions is too rigid to inspire support.

The entire underlined phrase is the subject of the verb *is.* But agreement between the subject and verb is determined by the headword of the phrase—the noun that the rest of the phrase modifies. The headword in the above phrase is *policy,* which is singular and takes the singular verb *is.*

5. A compound subject usually consists of two *coordinate,* or equal, sentence elements, rarely more. But whether two or more, they are always connected by *and, or,* or *nor.*

Since compound subjects are plural, they take a plural verb:

The president and vice-president never fly together.

Use a singular verb, however, when the two subjects are so closely related they are considered a single unit:

Assault and battery is a felony.

Note that the dividing line is not always clear. When in doubt, use what *sounds* right. A writer's natural sense of the sound of a sentence is the most dependable guide in the face of uncertainty. As the poet Robert Frost once said: "If the sound is right, the sense will take care of itself."

As between subject and verb, the question of agreement between pronoun and *antecedent* (the noun the pronoun stands in for) commonly concerns number. Pronouns, however, have more plural forms than do nouns, and they are more varied. Where most nouns form the plural by adding *-s*, most pronouns form the plural through radical internal inflections. The singular and plural forms of the most commonly used pronouns are shown in Table 3–2. These forms also are identified in terms of *person* (the speaker, the spoken to, the spoken about) and *case,* which indicates when a pronoun functions as a subject, object, or possessive.

TABLE 3-2. PRONOUNS

KIND OF PRONOUN	SINGULAR CASE			PLURAL CASE		
	Subj.	Obj.	Poss.	Sub.	Obj.	Poss.
PERSONAL 1st person	I	me	my, mine	we	us	our, ours
PERSONAL 2nd person	you	you	your, yours	you	you	your, yours
PERSONAL 3rd person	he she it	him her it	his her, hers its	they	them	their, their
RELATIVE	who	whom	whose	who	whom	whose
	that, which			that, which		
DEMONSTRATIVE	this, that			these, those		

Relative pronouns relate dependent clauses to other sentence elements. For example, in the sentence "Freud is the man who founded psychoanalysis," *who* is a relative pronoun. It relates the dependent clause, *who founded psychoanalysis,* to the main clause, *Freud is the man.*

Demonstrative pronouns limit nouns or stand in for

them. For example, in the sentence "These two engines are defective; these two are acceptable," *these,* used twice, is a demonstrative pronoun. The first *these* limits the defective engines to two—these two, not the others. The second *these* limits the acceptable engines in the same way and also stands in for the word *engines.*

Nouns of course no longer inflect for person, and they inflect for case only to show possession. To make a noun possessive, add an *apostrophe* (') and an *-s,* or just an apostrophe where the noun ends in *-s*: *the doctor's office; Dr. Jones' office; the doctors' offices. The doctor's office* indicates one doctor and one office, as does *Dr. Jones' office; the doctors' offices* indicates more than one doctor and more than one office. A third possibility, *the doctors' office,* indicates more than one doctor but only one office. Note that the plural and possessive noun forms are similar except for the apostrophe—which was created in the seventeenth century to distinguish between the two noun forms. An exception to this similarity are irregular plural nouns like *children* and *women,* which take an *-s* (and an apostrophe) *only* for the possessive. **(See also 5-3, p. 81.)**

Aside from the confusion that a missing apostrophe might cause with noun possessives, the question of agreement between nouns and pronouns presents little problem. Almost always, singular antecedents take singular pronouns, and plural antecedents take plural pronouns. Guided by our natural sense of English, we usually match the two automatically. But some pronouns present a problem: either they raise the issue of sexism or they produce a conflict between form and usage.

The issue of sexism arises when the masculine pronoun *he, him,* or *his* is used to represent both men and women. Although such usage has been traditional, increased awareness of its subtle sexism has triggered a search for a nonsexist alternative. Feminists have coined such pronouns as *co, tey, tem,* and *ter* to represent both sexes, but hardly anyone uses them. One alternative, of course, is to switch to the passive voice in such situations. Another is to use plural instead of singular pronouns and antecedents. For example, in the sentence

"A pilot must check his instruments before takeoff," where *pilot* may be a man or woman, change the pronoun and antecedent: *"Pilots* must check *their* instruments before takeoff." Plural pronouns have no *gender,* and so are not sexist. **(See also 3-4, p. 50.)**

The question is one of choices as much as of agreement. Whether to ignore or avoid subtle sexism is the first choice. The second one, if it arises, is between the passive voice and the plural pronoun (or even the feminist pronouns). The passive voice is safe ("Instruments must be checked by the pilot before takeoff") but weak. The plural pronoun is strong, but it is not always applicable. For example, in the sentence "Everything you write is used by *someone* to help *him* make a decision," changing the pronoun and antecedent to the plural would change the sense of the statement.

The conflict between form and usage is most evident in the pronoun *who*—or rather in two of its three case forms, *whom* and *whose.* The subjective case form is *who,* the objective form is *whom,* and the possessive form is *whose.* Prescriptive grammarians insist that *whom* should be used when it functions as an object: "For *whom* are you looking?" But the tendency in English usage is to shorten and simplify. Most people don't bother with the case distinction between *who* and *whom;* they simply say: *"Who* are you looking for?"

Whose, the possessive case of *who,* poses a different problem. Prescriptive grammarians maintain that *whose* should be used only when it refers to persons: "Fleming, *whose* discovery of penicillin was accidental . . ." With all other nouns, they insist, only the phrase *the . . . of which* is proper: "The report, *the* contents *of which* were misleading . . ." But most people use *whose* in all situations: "The report, *whose* contents were misleading . . ."

With little basis in usage, any form loses its vitality (as *whom* and *the . . . of which* have) and eventually withers away. That is the way in English: in a clash between form and usage, usage normally prevails. Were it not so, we might still be speaking and writing Old English. Instead of saying "It is me," or even "It is I," we might still be saying "Ic hyt eom," as the Anglo-Saxons did.

3-3 REPLACE STATIC VERB PHRASES WITH STRONG VERBS FOR MORE DIRECT, FORCEFUL SENTENCES.

The verb is the engine of the sentence: it generates power. Mired in a static phrase, however, it stalls—and so does the sentence.

We produce a static verb phrase when we convert a strong verb to its noun form and attach it to a weaker verb, usually as the object of the weaker verb. The habit carries over from our conversation, where we do this regularly. *Make a decision,* for example, is a static verb phrase. A strong verb, *decide,* has been changed to its noun form and attached to the weaker verb *make. Come to an agreement* is another example. *Agree* in its noun form has been attached to the weaker *come.*

Verbs like *make, come, give, take, have,* and *is* frequently turn up in static verb phrases. By themselves, these verbs perform important functions, but they lack the power to move a phrase. Some common static verb phrases and the verbs to replace them with are:

arrive at a compromise	compromise
comes in conflict with	conflicts
gives a performance	performs
has a weakening effect on	weakens
hold a meeting	meet
is beneficial to	benefits
make an improvement	improve
perform an operation	operate
put a stop to	stop
take into consideration	consider

Some verb phrases are necessary. *Have written,* for example, indicates tense; *was hired* indicates the passive voice; *may have* indicates the subjunctive mood. **(See also 3-4, p. 50, and Appendix 3, p. 190.)**

Static verb phrases, however, clutter up sentences, turn them in circles. As you edit, look for verbs like *make, take, come, give, have,* and *is.* Chances are they begin a verb phrase that can be replaced by a strong verb, one that conveys a vivid sense of the action it describes.

3-4 CHOOSE THE ACTIVE VOICE FOR CLARITY AND FORCE, THE PASSIVE VOICE TO SHIFT EMPHASIS OR CHANGE THE PACE.

The active voice is direct and forceful; the passive voice is indirect and weak. What causes the difference? The explanation lies in the relationship between the subject and verb of the sentence. In the active voice, the subject *acts,* makes something happen:

ACTION →

Kesecker processes Chicago orders every Monday.

The active voice here performs two functions. One, it clearly identifies the subject, the *actor:* Kesecker. Two, it generates movement in the sentence. Since the object of Kesecker's action *(Chicago orders)* follows the act *(processes)* in the sentence, the action flows forward from *Kesecker* to *orders.*

In the passive voice, the subject is *acted upon;* that is, something happens *to* it:

← ACTION

Chicago orders are processed every Monday.

Note that *process* becomes *are processed.* The mark of the passive voice is the linking of a form of the verb *be* to the past participle of the main verb, as in *are processed* or *were being processed,* and so on. In some tenses, a form of the verb *can, will, shall,* or *have* is also required. For example: *can be processed, have been processed, had been processed, will have been processed, should have been processed.*

The change from active to passive voice produces opposite effects. First, the subject of the sentence shifts: the actor, *Kesecker,* is replaced by the object of the action, *Chicago orders.* The writer now focuses on the orders and what happens to them rather than on the person who processes them. Second, Kesecker vanishes;

the actor becomes anonymous. We cannot tell who (or what) processes the orders. Third, the sentence winds down, turns in on itself. The subject of the sentence (*Chicago orders*) is also the object of the action; and since it precedes the act (*are processed*), the action now reflects backward.

The anonymity produced by the passive voice appeals to scientists who wish to remove the human element from scientific writing to reflect the objectivity of scientific observation. But objectivity in science, as in everything else, is relative. Physicist Werner Heisenberg's Uncertainty Principle, which showed that absolute statements in physics could be no more than relative probabilities, suggests that the human element cannot be removed from observation. Completely objective observation is impossible because the observer is human, and people remain subjective despite their best intentions.

The active voice, then, even the use of the first person *I*, reflects the reality of observation, including scientific observation. It also relieves what physicist P. W. Bridgeman calls the "coldly impersonal generality" of scientific writing. But the guardians of tradition resist change. Although Heisenberg's Principle has been accepted since 1927, most scientists still prefer the passive voice. So do most technical writers in business, industry, and government, for the passive voice lends to technical writing not only science's reputation for objectivity but its aura of authority. And in a pinch it provides the safety of ambiguity, the shelter of anonymity.

Uses of the Passive Voice

In all its uses, the passive remains a weak voice. It functions well, however, in the following situations:

1. *When the actor is unknown.* If the processor of the Chicago orders were unknown, you would have to write: "The Chicago orders are processed. . . ."
2. *When the object of the action is more important than*

the actor, for example, if the reader wants to know *when* the orders were processed instead of who processed them. If the orders are more important than Kesecker, write: "The Chicago orders are processed. . . ." To identify the actor (optional), add "by Kesecker."

3. *Where the active voice would be awkward,* as in the sentence: "Most paperbacks are bound by the perfect binding process."

4. *When the active voice would reflect subtle sexism,* as in the sentences: "After the typographer sets the type, he pulls galley proofs of it. First, he inks the type." In each sentence the pronoun *he* is meant to represent both men and women; but technically *he* excludes women. The pronouns *his* and *him* raise the same problem.

Where a pronoun is meant to represent both sexes, *he or she, him or her, his or hers,* and *himself and herself* are often substituted for the male equivalent. But these are awkward, as are the abridged forms, *he/she, him/her, his/hers,* and *himself/herself.*

Feminists suggest several alternatives. One is to substitute *co* for *he, she, him, her, his,* and *hers,* adding an apostrophe and an "s" to indicate possession *(co's),* and to use *coself* for *himself* or *herself.* Another is to use *tey* for *he* or *she, tem* for *him* or *her, ter* for *his* or *her,* and *temself* for *himself* or *herself. Tey, tem, ter,* and *temself* have been approved by the American Psychological Association. But the shapers of the language—that is, the ordinary speakers of it—show no inclination to adopt either alternative.

A simpler solution is to change the antecedent *(typographer)* from singular to plural: "After the typographers set the type, *they* pull galley proofs of it. First, *they* ink the type." A plural antecedent takes a plural pronoun *(they, them, their, themselves),* and plural pronouns include both sexes.

The simplest solution is to switch to the passive voice: "After the type is set, galley proofs of it are pulled. First, the type is inked."

Shifting voice in one sentence but not in the other is another alternative: "After the type is set, galley proofs of it are pulled. First, the typographer inks the type." This generates movement and introduces the actor.

Take care, however, in shifting voices in mid-sentence: "After the typographer sets the type, galley proofs of it are pulled." This sentence starts strong but weakens at the end because the main clause is in the passive voice. A more effective version is: "After the type is set, the typographer pulls galley proofs of it." With the dependent clause in the passive voice and the main clause in the active voice, the sentence starts slowly and gains momentum toward the end.

5. *When a change of pace is timely.* The passive voice adds variety by changing the pace, slowing it down. In long passages written in the active voice, the passive voice can even be used for emphasis because it will stand out by contrast.

3-5 AVOID SENTENCE STRUCTURE THAT DISTORTS OR DESTROYS MEANING.

The placement of modifiers in the sentence is the key here. A carelessly placed modifier can completely change the intended meaning of a sentence. For example, note the difference in meaning between the two versions of one sentence:

> The instructor slowly explained how to operate the machine.
>
> The instructor explained how to operate the machine slowly.

In the first version, *slowly* modifies the verb *explained:* the instructor explains slowly. In the second version, *slowly* modifies the verbal *operate*; here the machine operates slowly. Quite a difference.

To say what you mean, place the modifier close to the word it modifies. Usually, a modifier is placed before the word it modifies—*red* dye or *slowly* explains. The word *only*, however, is an exception in certain contexts. For example, in the sentence "The inspector only found three defective switches," *only* modifies the noun phrase *three defective switches,* but is commonly placed before instead of after the verb *found.* This placement subtly affects the message of the sentence. Placed after the verb *(found only three defective switches), only* implies that other switches, acceptable switches, were also to be found. Placed before the verb *(only found three defective switches), only* implies that all there was to be found were defective switches. Whichever way the sentence is written, however, usage has accustomed most readers to see in it the first message.

Place *only* carefully, however, when its position affects meaning:

Andrews can only write when she is in her office.

Andrews can write only when she is in her office.

In the first version, the only thing Andrews can do in her office is write—she can't read, for example, or conduct an interview. In the second version, the office is the only place where Andrews can write. The two versions thus convey quite different meanings.

Two specific placements disputed by prescriptive and descriptive grammarians deserve mention. These are the controversies concerning ending a sentence with a preposition and splitting an infinitive:

In the actual bidding, we had little competition to contend with.

Local 651 is believed to strongly favor a strike.

The first sentence ends with the preposition *with,* and in the second sentence the adverb *strongly* splits the infinitive *to favor.* Both usages have traditionally been discouraged by prescriptive grammarians. But two points are worth noting. One, usage favors and meaning depends on such constructions in certain sentence patterns. The price of avoiding them may be an awkward or distorted sentence. Two, both bans are based on Latin syntax and have no structural relevance to English, which is a Germanic language.

3-6 VARY THE LENGTH AND STRUCTURE OF SENTENCES TO AVOID MONOTONY AND MAKE THEM MORE EFFECTIVE.

There is no ideal sentence length or structure, only degrees of effectiveness. How effective a sentence is—that is, how well it does its job—depends on how the writer weighs the four factors that affect all technical writing. These are: (1) the audience; (2) the purpose—whether it is to inform, instruct, or persuade; (3) the number, complexity, and comparative importance of the ideas; and (4) the context, especially the preceding and succeeding sentences. Weighing these factors, the writer aims to produce clear, lively sentences.

Audience

The main question where audience is concerned is its reading level. An audience with a low reading level—technical or general—requires simpler sentences and more definitions of terms than one with a high reading level. So write mostly short sentences, and define terms separately.

Studies of reading levels indicate that writing is easy to read when the average length of sentences is 14 words or less. Where the reading level is high, writing with an average of 17 to 20 words per sentence is effective. The average sentence length in this book is 15 words.

Purpose

Where purpose is concerned, it is harder to persuade than to inform a reader, and hardest to instruct one. Writing that informs must be clear, and writing that persuades must be clear and logical (except in advertising). But writing that instructs must be clear, logical, and stimulating—and helpful.

For clarity, mix long with short sentences, simple with complex ones. For clarity and logic, lean more toward simple sentences. And to make writing stimulating as well as clear and logical, vary sentence patterns as well as sentence length. For example, use S-P, M-S-P, even P-S sentence patterns; make the modifier a word, phrase, or clause; use prepositional, verbal, or noun phrases. **(See 3-1, p. 33, and below for more on phrases, clauses, and sentence patterns.)**

Ideas

There is no arbitrary limit on the number of ideas that will work in a sentence. How many ideas a sentence can carry depends not only on their number but also on their complexity, their relationships, and the writer's skill in structuring the sentence. The structure of the sentence is directly related to the relationship between the ideas.

Ideas that carry equal weight are *coordinate;* where the weight is unequal, one idea must be *subordinate.* Coordination and subordination—that is, equality and inequality—are most effectively expressed through sentence structure. **(For the uses of coordination and subordination in the shaping of paragraphs, see 4-3, p. 66.)** In grammar, the three most common sentence structures are called *simple, compound,* and *complex.* The names, however, are misleading; they describe only the nature of the structure, for a simple sentence can express a complex idea, and vice versa.

The structures themselves are easily distinguished. A simple sentence contains only one independent clause; a compound sentence contains two or more; a complex sentence contains at least one independent clause and any number of dependent (subordinate) clauses. **(See also 3-1, p. 33.)** An example of each follows (the dependent clause is underlined):

Simple: This is a simple sentence.

Compound: A compound sentence expresses coordination well, and a complex sentence expresses subordination well.

Complex: When the weight of two ideas is unequal, one idea must be subordinate to the other.

Few writers have trouble with compound sentences, which after all are merely simple sentences connected by a coordinating conjunction. Complex sentences, however, often pose a problem, perhaps because writers are unsure of when to subordinate one idea to another. The following relationships usually involve subordination: (a) those of cause-and-effect; (b) those that are conditional—that is, possibilities or probabilities that depend on something, someone, or some time; and (c) those in which the ideas are contrasted to emphasize the importance of one. The three kinds of subordination are illustrated below:

Cause-and-effect: The power went off because a central generator failed.

Conditional: If the economy recovers, profits will rise.

Contrast: Although the design was changed, valve performance remains unacceptable.

Note that the conjunctions used in each situation indicate the nature of the relationship.

In complex sentences, the subordinate, or dependent, clause is connected to the independent clause by a *subordinating* conjunction. A list of subordinating conjunctions appears in Table 3-3. **(See Table 3-1 on page 41 for a list of conjunctions and conjunctive adverbs.)**

TABLE 3-3. SUBORDINATING CONJUNCTIONS

after	before	like	so (that)
although	even though	more than	than
as	fewer than	no matter how	though
as if	if	now that	unless
as long as	in case	once	until
as soon as	in order that	provided (that)	when
as though	in that	since	where
because	less than	so long as	while

Context

The immediate context of a sentence are the sentences before and after it; the larger context is the paragraph. In the immediate context, sentence variation is largely a matter of preference. But in a long paragraph, some variation in sentence length and structure is essential to avoid monotony.

3-7 USE PARALLEL STRUCTURE TO STRESS THE SIMILARITY BETWEEN LIKE IDEAS, OBJECTS, OR PROCESSES.

A series of two or more similar words, phrases, or clauses with identical grammatical structure are said to be *parallel.* Usually the parallel elements are connected by a conjunction such as *and, but, or, nor,* or *yet.* Some examples of parallel structure are: "trial and error," "with the education but without the knowledge," "When in the minority, talk; when in the majority, vote; when in doubt, pray."

Parallelism coordinates the parallel elements and emphasizes the connection between them. It also creates a consistent, distinctive rhythm, which is part of its power. To remove a word or comma from the parallel structures above—for example, the *the* before *education* or *knowledge,* the comma before *talk* or *vote*—would destroy the parallelism and therefore the rhythm of the structure. Rhythm and structure: together they produce the clear, concise sentence that is valued, but rarely found, in technical writing. Note the differences between the sentences below:

In compression the rod breaks at a 45-degree angle. The tensile fracture is perpendicular.	In compression the rod fractures at a 45-degree angle; in tension it fractures at a 90-degree angle.

Each version is grammatically correct. But four revisions in the version on the right created parallel structure, adding clarity and force to the sentence. The first, and decisive, revision changed the pattern of the second sentence from S-P to M-S-P in order to parallel the pattern of the first sentence *(In compression*... and *in tension . . .).* This coordinated the two operations and created a consistent rhythm. The second revision was to connect both sentences with a semicolon to emphasize their similarity. The third was to replace *breaks* with *fractures* in the first clause. *Fractures* in a more effective term in this context and is parallel with *fractures* in the second clause. Finally, *perpendicular* was changed to *90-degree angle* to make it parallel with *45-degree angle.*

3-8 RECAST A SENTENCE THAT SEEMS WEAK OR AWKWARD; IF IT STILL FALTERS, QUESTION ITS RELEVANCE.

In one respect, sentences are like houses: some can be repaired, others must be razed and completely rebuilt. If rebuilding fails, reconsider what you are trying to say. Is it necessary? Is it important? Is it relevant? Strike it out if the answer is *no* to any of these questions.

Often the concept behind the sentence has not been thought through. If it hasn't, analyze it further. Do not, however, hang on to an idea or sentence simply because it is clever. Such sentences are like "fool's gold," and about as useful.

EXERCISES

In the paragraphs below—

1. Identify the fragment sentence and the run-on sentence, and rewrite them as complete sentences.

2. Identify a simple sentence; a compound sentence; a complex sentence; a subordinate clause; a phrase.

3. Identify an S-P sentence pattern; an M-S-P sentence pattern.

(1) A probability of 0 makes an event impossible; a probability of 1 makes it certain. (2) All other probabilities are expressed in fractions falling between 0 and 1. (3) In a probability of one-half, the chances are 50-50, or even. (4) A probability of one-fourth is less than an even chance the event has 1 chance in 4 of happening. (5) In other words, the odds are 3 to 1 that the event won't happen.

(6) Every chance event is independent of every other chance event. (7) For example, if a coin is tossed 10,000 times, it will come up heads and tails about the same number of times, which is what can be expected on a 50-50 chance. (8) But in a short run, the coin may go against the odds and come up heads. (9) Or tails 10 or 20 times in succession. (10) When this happens, many gamblers bet against the streak continuing; they feel the streak must even out. (11) This is called the "gambler's fallacy" because the odds even out over the long run, not the short run.

4. Rewrite the following sentences, making any necessary changes to remove ambiguity, clarify meaning, or adjust dangling sentence elements:

After checking with the receiving department, it appears that the shipment of March 15 never arrived.

A plastic finish is sprayed on each side of the board and hung to dry for three hours.

The following is my assessment of the Hewlett-Packard Computerized Arrhythmia Detection System that you requested.

After 30 seconds in the fix, examine the test print under the light.

5. Rewrite the following sentences, making any necessary changes to provide parallel structure; adjust agreement between subject and verb, or between pronoun and antecedent; or avoid subtle sexism:

The ratio of the number of turns in the coils in the transformer determine whether it steps up or steps down the voltage.

Cutting fluids cool and lubricate the cutter and the workpiece, preserving tools, increasing production, and the reduction of costs.

The astronaut training program is physically strenuous, emotionally demanding, and challenges the intellect.

If a student is unable to complete the course, he may apply for an "incomplete" grade.

The program included a class in R.F. theory and one in thermodynamics. It was a valuable class and satisfied a graduation requirement.

6. In the following passage, circle each static verb phrase and replace it with a strong verb:

County and union officials will hold a meeting tomorrow to take into consideration the arbitrator's latest proposal for putting an end to the strike that has had such a crippling effect on public transportation. If the proposal is beneficial to the employees and does not come in conflict with state law, both sides will probably come to an agreement.

7. Rewrite the following paragraph in the active voice and compare the two versions. Then rewrite the paragraph using both the active and passive voice, each where it seems more effective.

High concentrations of protein are found in soybeans, peanuts, cottonseed, and sesame. The advantage of the soybean is that it can be ground into flour economically on the spot, without skilled labor or electric power. With a simple hand process that has been developed by the U.S. Department of Agriculture, 300 pounds of soybean flour can be produced by 5 men working an 8-hour day, enough to supply the daily protein requirements of 1600 adults.

The visible aspects of organization are consistency, logic, sequence, relevance, and balance, each of which is discussed in this section. But behind all these lies the ability to perceive the relationships among the parts of a whole. That is the key. For example, the following 15 numbers can be arranged according to at least 8 relationships among them:

5, 14, 6, 28, 40, 56, 10, 49, 25, 7, 9, 20, 42, 3, 77

The 8 relationships are: (1) odd-even; (2) low-to-high order; (3) high-to-low order; (4) higher and lower than the median, 20; (5) higher and lower than the average, 26; (6) multiples of 3, 5, or 7; (7) single versus double digit numbers; and (8) alphabetical order of written equivalents.

Order, the essence of organization, is here imposed on the numbers by each of the eight patterns. It doesn't matter that the patterns are different. In writing, order is whatever pattern enables us to function effectively—that is, write clearly and coherently.

4 ORGANIZATION

4-1 TO ORGANIZE THE MATERIAL OF A SUBJECT, FIRST BREAK IT DOWN INTO ITS COMPONENT ASPECTS.

Analysis, the breaking down of things, and classification, the separating of them into consistently uniform groups, are basic organizing procedures. So is synthesis, the combining of things. All three depend on the writer's ability to perceive the relationships among the parts of the whole—what is similar, different, coordinate, subordinate, and so on.

The subject of minerals, for example, can be analyzed and the results classified on the basis of at least seven kinds of relationships among them. These are: (1) geographic location, (2) geological age, (3) method of formation—igneous, sedimentary, or metamorphic, (4) physical properties—color, hardness, specific gravity, fracture, and so on, (5) crystal systems—cubic, tetragonal, hexagonal, and so on, (6) chemical structure—elements, sulphides, oxides, halides, and so on, and (7) individual minerals—acanthite, acmite, actinolite, adamite, and so on.

The classification a writer uses depends upon the nature and purpose of the report or paper. For example, an oil geologist would classify minerals differently than a gemologist (gem expert) because their interests and purposes differ.

Once analysis and classification are completed, the next step is to synthesize the resultant groups into a logical, consistent whole. The process is not unlike dismantling a puzzle and rearranging the pieces in a straight—that is, sequential—line, as in Figure 4-1.

FIGURE 4-1. ORGANIZING RANDOM MATERIAL

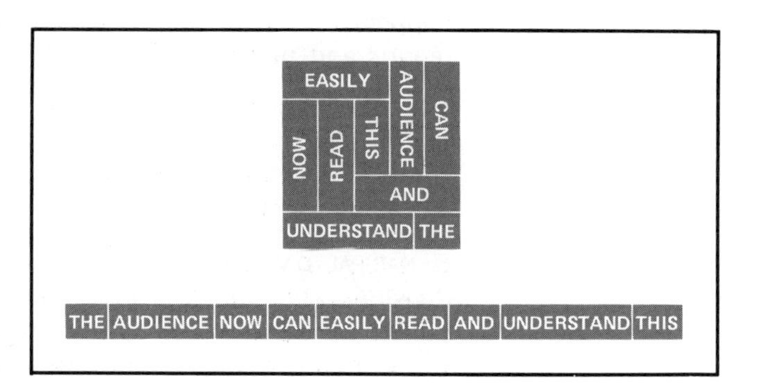

In the organization of any piece of writing, some aspects of the subject are more significant or more dramatic than others. For maximum effect, open with the most significant and close with the next most significant or vice versa. Use the middle for development. A forceful opening engages the reader and sets up the further development of the opening statement, argument, or problem. And a forceful ending leaves the reader with a memorable impression.

This pattern of emphasis also works well in a paragraph. **(See also 4-3, p. 66.)**

4-2 TO ORGANIZE A REPORT OR PAPER, CHOOSE A SUITABLE APPROACH AND MAKE AN OUTLINE THAT IMPLEMENTS IT.

The approach to a subject depends on the nature of the material, the writer's purpose, and the audience's needs. Purpose and audience are discusssed in Section 1-3; the focus here is on the material. Five approaches often used, depending on the nature of the material, are described below:

1. **PROBLEM-AND-SOLUTION** Based on a problem and a projected solution to it. Often used in proposals and feasibility and complaint reports. It includes an explanation of the method of solution.

2. **CAUSE-AND-EFFECT** Based on causal relationship—A is caused by B or A is the result of B. Also useful in proposals and feasibility and complaint reports.

3. **CHRONOLOGICAL** Based on time sequence—first this happened, then this, and so on. Effective in progress reports and in trip and conference reports.

4. **SPATIAL** Based on area or location—from top to bottom, left to right, north-east-south-west around the compass, and so on. Effective in survey reports and in trip and conference reports.

5. **NATURAL DIVISION** Based on natural or established divisions—14 branch stores, 4 departments, geology-biology, government-nongovernment, and so on. Useful in survey or status reports, or in any report or paper based on comparisons.

These approaches often are used together in one report or paper. So are the *deductive* and *inductive* methods, which lead, respectively, from the general to the specific, and from the specific to the general. **(See also 1-6, p. 13.)** Where the material involves controversy, however, the inductive method is more persuasive. Deduction begins with a statement that is then supported by evidence. Psychologically, this sequence risks alienating a reader who, resisting the statement, may view even the strongest subsequent evidence negatively. Such alienation is avoided with the inductive method, where the evidence gradually leads to a concluding statement and can be accepted at face value by even a doubting reader.

The use of the cause-and-effect and deductive and inductive approaches requires an understanding of the difference between fact, inference, and opinion. A fact has been proven true by conclusive evidence. For example, it is a fact that the earth revolves around the sun. An inference is based on evidence that is reliable but not sufficient to be conclusive. We can infer from the evidence, for example, that the continents were once a single land mass. An opinion, however, lacks such a verifiable frame of reference. It may even be based simply on emotional attitudes (one of the Latin roots of *opinion* is *optare,* "to select, desire"). That earth has been visited by intelligent creatures from outer space, for example, is an opinion.

It is worth noting, however, that the passage of time or a change in perspective may alter the way we perceive anything. For example, though today it is a fact that the earth revolves around the sun, less than 500 years ago it was a "fact" that the sun revolved around the earth. Today, strong evidence supports the inference that the continents were once a single land mass. But 50 years ago continental drift was only an opinion. Finally, with the increasing evidence of the existence of UFOs, the current opinion that earth has been visited by alien creatures from space may become tomorrow's inference.

The final organizational step is a rough outline, the basic blueprint that brings all the material of the report or paper into focus. Such an outline should indicate to the writer a direct line of attack—what to connect, where to begin, what direction to take, what to end with. At the same time it should clarify what to emphasize, what to develop, what to drop. An effective outline will even expose what may be missing in the coverage of the subject. **(For further discussion of outlines, see 8-1, p. 156; 8-2, p. 158; and 8-3, p. 177.)**

4-3 THE BASIC UNIT OF ORGANIZATION IS THE PARAGRAPH.

Paragraphs act as a visual aid and mental tonic. They break up what would otherwise be a formidable mass of words, and they allow the reader a brief rest stop—a breather.

Structurally, the paragraph is both the sentence writ large and the entire report or paper writ small. Like a sentence, the paragraph is usually complete in itself, and it must fit into a larger order. And like the entire report or paper, the paragraph is a collection of related ideas, logically and consistently organized.

The question is how to organize it. Traditionally, the paragraph is organized around a *topic sentence,* a statement of the governing idea of the paragraph and usually the lead sentence. More recently, the teacher and writer Francis Christensen has suggested viewing the paragraph as "a group of sentences related to one another by coordination and subordination." The two viewpoints complement more than contradict each other.

The traditional approach works well if we look at the topic sentence as raising a question, or questions, that the rest of the paragraph will answer. The body of the paragraph then becomes supporting detail that answers the question or questions *where? when? how? why?* or *to what extent?* This detail can be developed through explanation, example, or comparison.

For example, in the first paragraph of this subsection, the first sentence *(Paragraphs act...)* is the topic sentence, and the second sentence answers the question *how?* The lead sentence of the second paragraph

(Structurally . . .) is also the topic sentence. There the answer *how* is developed through comparisons. A paragraph is compared first to a sentence, then to a report or paper.

These techniques of development can be combined, as they are, for example, in the third paragraph of this subsection *(The question is . . .).* The topic sentence, which poses the question of how to organize a paragraph, is followed by an explanation of two approaches. And the paragraph closes with a comparison of the two.

Note also that the three paragraphs vary in length. There is no standard paragraph length, only the length appropriate for the subject, the writer's purpose, and the audience. Most of the paragraphs in this book are short because the subject is difficult, the purpose is to instruct, and the audience's reading level varies. Difficult concepts are easier to understand in small units; the frequent rest stops also help keep the reader fresh. In paragraphing, as in all aspects of technical writing, it pays to fit the form to the function rather than vice versa.

Christensen views the topic sentence not as a question but as the sentence on which the other sentences in the paragraph depend. It is the sentence whose statement is *developed* by the sentences that follow—that is, confirmed, explained, detailed, or illustrated by them. The development itself is always subordinate to the topic sentence. (The development of something cannot be equal to the original; equality, or coordination, requires likeness to the original—for example, a restatement.) The subordinate sentences themselves, however, may be coordinate with each other, as in two details or three examples.

Subordination, then, adds depth, which is what development does (and what the prefix *sub* implies). And coordination adds emphasis, as in a restatement (or four coordinate examples). The two complement each other and hold the paragraph together. Each sentence is either coordinate with any sentence above it or subordinate to the one immediately above it. A sentence that is neither, Christensen says, indicates that the paragraph "has begun to drift from its moorings, or the writer has unwittingly begun a new paragraph."

Applying this perspective to the first three paragraphs in this subsection, we find parallels to the traditional approach. In the first paragraph *(Paragraphs act...)* the second sentence explains the first one and is subordinate to it. This pattern is repeated in the second paragraph *(Structurally...)*: the lead sentence is explained (through comparisons) by the next two sentences, which are subordinate to the lead and coordinate with each other. The pattern is similar in the third paragraph *(The question is...)*: the lead sentence is explained, in one way or another, by the next three sentences, and all are subordinate to the first and coordinate with each other. **(See also 7-4, p. 146.)**

If there is any significant difference between the traditional approach to paragraphs and Christensen's, it is in our familiarity with the traditional. More important than any differences between them is their common focus on development. Without development, the paragraph is just a collection of sentences. With it, the whole becomes more than the sum of its parts.

EXERCISES

Using the data below—

1. Identify five statements of fact.
2. Identify one statement of inference.
3. Identify one statement of opinion.
4. Organize the statements into an outline for a report that would review the water supply situation in Sonoma County from July 1975 to January 1979. Use one of the five organizational approaches discussed in this section.
5. Take the same material and reorganize it, using another of the five organizational approaches.
6. Write an opening paragraph organized around a topic sentence.

7. Write an opening paragraph based on coordinate and subordinate relationships.

(1) Rainfall for the first 6 months of the 1978 season was 3.50 inches.

(2) Drought emergency restrictions were lifted January 1978.

(3) Rainfall for the first 6 months of the 1975 season was 7.11 inches.

(4) The rainy season runs from July through June.

(5) Rainfall for 1975–76 was 14.47 inches.

(6) Rainfall for the first 6 months of the 1977 season was 16.91 inches.

(7) Ground water level dropped 25 percent below normal in the period under study.

(8) Normal rainfall is 30.00 inches/year.

(9) Drought emergency restrictions went into effect in June 1977.

(10) Rainfall for 1977–78 was 44.11 inches.

(11) Normal rainfall for the first 6 months of the season is 11.00 inches.

(12) The population of Sonoma County is increasing by 5 percent annually.

(13) Rainfall for 1976–77 was 12.22 inches.

(14) Rainfall for the first 6 months of the 1976 season was 5.80 inches.

8. Exchange reports with another student and analyze the organization of that student's report; that is, make an outline of it. Point out the use of the deductive or inductive method, chronological sequence, natural division, and other approaches to organization.

9. Using the diagram on page 70, write an accident report describing the collision between cars A and B. Be sure to separate opinion from inference, and both from fact. For information on right-of-way, consult the state motor vehicle department.

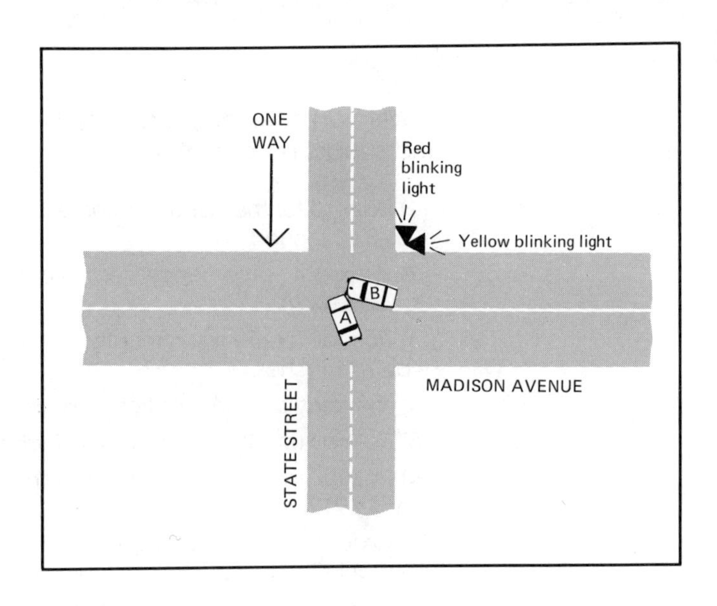
ONE
WAY
Red
blinking
light
Yellow blinking light
A
B
MADISON AVENUE
STATE STREET

Punctuation is the written equivalent of the rhythm and emphasis in our speech. We drop our voice when we end a sentence and place a period after one when we write it. The question mark substitutes for the rise of our voice in a question, the comma for short pauses, the semicolon for longer ones. The exclamation point indicates emphasis, surprise, joy, or shock. And so on.

In the long history of writing, however, punctuation appears relatively late. The earliest punctuation mark was a dot used by Greek scribes in the third century B.C. to indicate a break in a sentence. Placed high, low, or midway on the line of script, it became the ancestor of the period, comma, colon, and semicolon. But these and other marks were used inconsistently, and by the eighteenth century, English punctuation was as confusing as English spelling. Of the 47 sentences of the original Declaration of Independence, for example, 35 end with a comma plus dash (,—), a colon plus dash (:—) or a period plus dash (.—).

None of these combinations are used today. Like modern American English, modern punctuation is becoming simpler. We tend to punctuate less, to use only what is necessary when it is necessary. Such usage reflects the principle that the function of language—what it does and how well it does it—counts more than the form it takes. This applies particularly to technical writing.

Punctuation's first function is to clarify what is written; its second is to create rhythm—pace. If it fulfills the first, it has done its job; if it fulfills the second, it has done its job with style.

5-1 THE FOUR MAIN USES OF PUNCTUATION ARE TO END SENTENCES AND TO INTRODUCE, SEPARATE, OR ENCLOSE LETTERS, WORDS, PHRASES, OR CLAUSES.

Of the 14 common punctuation marks, 11 have only one use. The 3 with multiple uses are ellipses (. . .), the comma (,), and the dash (—). Ellipses have 2 uses, the comma has 3, and the dash, the most versatile punctuation mark, has 4. Keep in mind, however, that any classification according to utility is arbitrary, and this one emphasizes common usage. For example, the colon, which introduces words, phrases, or clauses, is listed in Table 5-1 only under the category INTRODUCE. Technically, however, it can end one statement while introducing another (even while it separates the two): "The project was canceled for only one reason: it cost too much." But the colon's basic use is to introduce elements, not to end or separate them.

The 14 punctuation marks are: the period (.), question mark (?), exclamation point (!), ellipses (. . .), dash (—), comma (,), colon (:), semicolon (;), hyphen (-), apostrophe ('), slash (/), quote marks (" "), parentheses (), and brackets []. They are listed by uses in Table 5-1.

TABLE 5-1. USES OF PUNCTUATION MARKS

END	INTRODUCE	SEPARATE	ENCLOSE
Period	Comma	Comma	Comma
Exclamation point	Colon	Semicolon	Quote marks
		Hyphen	Parentheses
Question mark		Apostrophe	Brackets
Ellipses	Ellipses	Slash	
Dash	Dash	Dash	Dash

Each of the above is examined separately in this section except the question mark, the exclamation point, ellipses, and the slash, which are discussed below.

Question mark

The question mark raises only two problems—its use with quote marks and its use in an indirect question. When using a question mark with quote marks, place it inside the quote marks if it is part of the quote, outside if it is not. (The same applies to exclamation points used with quote marks—**see also 5-8**.) For example:

> The arbitrator asked, "What is the company's position?"
>
> Did the negotiator really say, "Take it or leave it"?

A direct question, of course, ends with a question mark: "What is the union's position?" But omit the question mark in an indirect question: "The arbitrator asked what the union's position was."

Exclamation point and italics

The exclamation point indicates emphasis except when used inside parentheses (!); then it implies irony. The only point to make about the exclamation point is to use it sparingly and to avoid ending a sentence with two (!!) or more. *Italics*, a typographical device, also show emphasis. (In handwritten or typewritten drafts, the equivalent of italics is a line drawn under the word or phrase to be emphasized.) Italics also are used (especially in this book) to refer to a word as such, and with foreign words, as in the following sentence: "*Technical* is derived from the Greek *technē* and *tektōn*." Also italicized are titles of books, magazines, plays, newspapers, movies, operas, symphonies, long poems, and legal cases (*Bakke* v. *University of California*).

Ellipses

Ellipsis points (*ellipses* in the plural) indicate omission. The omission may be in the beginning (fade-in), end (fade-out), or middle of a passage. (Ellipses may also indicate that the missing word or passage is illegible.)

Use either three (. . .) or four (. . . .) ellipses, no more. When the omission follows an incompleted sentence, use three. When it follows a completed sentence, use four—the fourth being the period ordinarily placed at the end of a sentence. If the omission occurs *before* the sentence you wish to start with, use three ellipses.

The use of ellipses, omitting portions of this paragraph, is illustrated below:

FADE-IN AND FADE-OUT: "...When the omission follows an incompleted sentence, use three; when it follows a completed sentence use four...."

OMISSION, INCOMPLETE SENTENCE: "Use either three (...) or four (....) ellipses..."

OMISSION, COMPLETE SENTENCE: "Use either three (...) or four (....) ellipses, no more....When it follows a completed sentence, use four—the fourth being the period ordinarily placed at the end of the sentence."

Note that ellipses go inside quote marks. Also, a question mark or an exclamation point—but no other punctuation—may be used with ellipses. If one is used, place it either before or after the ellipses.

Slash

Also called a bar, the technical name for the slash is *virgule,* after the Latin *virgula,* which means "little rod." Its most common use, the diagonal between "and/or," seems awkward. As used in constructions like "his/her," the slash itself means "and or"; but the constructions remain awkward. More practical as well as more acceptable are its uses as a substitute for "per" ($5/ton), as a divider in dates (9/17/78), in typing fractions (¾, ⅞), and in abbreviating the expression "in care of" (c/o). It is also used to separate consecutive lines of poetry that are set on the same line, as here in Emerson's "Things are in the saddle/And ride mankind," from his *Ode.*

5-2 THE COMMA SEPARATES, ENCLOSES, AND INTRODUCES; BUT ITS MAIN USE IS TO SEPARATE.

The word *comma* comes from the Greek word for "clause," *komma,* which in turn is derived from the Greek *koptein,* "to cut off." It is the most frequently used punctuation mark, and the most confusing, since its uses overlap. An entire section could be devoted to the comma, but here we will examine its common, and confusing, uses. In these respects keep in mind two points. One, the comma should always clarify, never confuse. Two, its spoken equivalent is the short pause.

Separate

Here the comma is used to avoid confusion or ambiguity and to express the desired pauses in a sentence. The four common uses of the comma as a separator are: (1) in a series of three or more words, phrases, or clauses; (2) with a conjunction between independent clauses; (3) with an opening, or closing, word, phrase, or clause of a sentence; (4) to omit or emphasize words.

1. In a series of three or more words, phrases, or clauses, place a comma before the *and* or *or*: "The comma encloses, introduces, and separates." Also use a comma if it is needed to clarify the sentence, or to indicate the separation of sentence elements, as in: "The subdivisions of X, Y, and Z, whose values are known, act as a check against overdevelopment." Without the comma before *and, Y* and *Z* could be interpreted as being subdivisions of *X*. The comma makes clear that *Y* and *Z* are separate and that there are three equal subdivisions: *X, Y,* and *Z*.

2. To separate independent clauses, use a comma with a conjunction. If the clauses are short, however, the comma *or* the conjunction (but not both) can be omitted. For example:

Business increased and so did profits.

Business increased, so did profits.

With longer clauses the alternative to the comma and conjunction is a period, or a semicolon if the clauses are closely related. (**See also 3-1, p. 33.**) The comma indicates a slight pause, the semicolon a longer pause (and a close relationship), and the period a full stop. (**See also 5-5, p. 84.**) The difference lies in effect, as the following three versions of the same sentence indicate:

With public sentiment nothing can fail, and without it nothing can succeed.

With public sentiment nothing can fail; without it, nothing can succeed.

With public sentiment nothing can fail. Without it, nothing can succeed.

The first version is the most fluid, the third the most clipped. Somewhere between them lies the second version, which was written by Abraham Lincoln.

3. Use a comma to separate an opening—or closing—word, phrase, or clause from the main body of the sentence under the following conditions: when the opening element precedes the subject of the sentence and omitting the comma would invite confusion; when the closing element follows a distinct pause. For example:

Before leaving, employees should turn out all lights.

The nation is in a depression, not a recession.

In the first sentence the phrase *Before leaving* precedes the subject (*employees*) and modifies it. Not all opening phrases, however, require a comma. The trend is to omit the opening comma—as it was omitted in the first sentence of this paragraph—when it is not needed to avoid confusion.

In the second sentence the phrase *not a recession* follows a distinct, natural pause. Note that replacing the comma with a dash would add emphasis to the statement.

4. The comma is often used to indicate omission in a series:

The first test uncovered 12 problem areas; the second, 9; the third, 1.

The repetitive phrases *test uncovered* and *problem areas* have been omitted in the second and third parts of the series. Here even the commas can be omitted without affecting clarity; the choice lies with the writer. If the commas were omitted, however, the semicolons could then be changed to commas. The total effect would be a more informal, swiftly moving sentence.

As for emphasis, the main emphatic devices are the dash, the exclamation point, and italics. But only the comma emphasizes subtly, even ironically, as in: "The cost-of-living index rose, again." Note that a dash would add emphasis and substitute disapproval for irony.

A final word about the comma: in addition to the uses described above, it is used to separate sentence elements whenever clarity demands it. The comma also can be used to separate connected elements for emphasis rather than for clarity. Note the sentence below:

> The comma is the most frequently used punctuation mark and the most confusing, since its uses overlap.

The sentence is clear, but a comma after *mark* would stress the confusion. Dashes after *mark* and *confusing* would emphasize the confusion even more.

Enclose

The comma encloses a word, phrase, or clause that interrupts the main idea of a sentence or is parenthetical to it. (**See also 5-7, p. 89.**)

An interruption is easily recognized; the flow of the sentence clearly changes: "One more heart attack, no matter how serious, may be your last."

A parenthetical element, however, sometimes takes commas, sometimes does not. It depends on its relationship to the sentence. In grammatical terms this relationship is expressed as *restrictive* or *nonrestrictive,* referring to the effect of the parenthetical element on the word or words it modifies. If it is restrictive—that is, necessary to the sense of the sentence—use *no* commas; if it is nonrestrictive, use commas. Put more simply: use commas when a word, phrase, or clause adds information to a sentence but *can be omitted* without destroying the sense of the sentence. The two sentences below illustrate the difference.

Restrictive:	Cars that have defective gas tanks are safety hazards.
Nonrestrictive:	Production of the defective gas tanks, which caused several deaths, has been discontinued.

In the first sentence the clause *that have defective gas tanks* cannot be omitted without destroying the sense of the sentence. To omit it would be to say that cars—all cars—are safety hazards. But the sense of the sentence is that only cars with defective gas tanks are safety hazards. So the clause *that have defective gas tanks* limits or *restricts* the kind of car that is a safety hazard, and commas are unnecessary.

In the second sentence the clause *which caused several deaths* adds information about the defective tanks. But it can be omitted without destroying the sense of the sentence, which is that the production of defective tanks has been discontinued. The clause is *nonrestrictive*—it says production of *all* defective tanks has been discontinued, not just production of defective tanks that caused fatalities, and so commas are necessary.

Note that the clause in sentence two also can be enclosed by dashes or parentheses. Since the dash adds emphasis, the use of dashes would play up the deaths. And since parentheses understate, the use of them would play down the deaths. Comma, which fall between dashes and parentheses in emphasis, simply state that the deaths occurred. All these options are grammatically correct; the choice of emphasis lies with the writer.

Enclosures require two commas. But when a nonrestrictive element ends a sentence, use only one comma:

The deaths caused by the defective gas tanks led to legal action, which is always costly.

Here the comma expands the sense of the sentence,

which is that the deaths caused by defective gas tanks led to legal action. The clause *which is always costly* is additional information that can be dropped without destroying the sense of the sentence. Nonrestrictive, the clause says that *all* legal action is always costly, not only legal action based on defective gas tanks.

Also enclosed are conjunctive adverbs (*however, also, therefore,* and so on) and expressions like *of course.* For example:

> Any pay increase, however, must be geared to the cost-of-living index. That will mean, of course, a smaller raise.

Not every *of course* requires commas. Whether or not commas are used in these instances depends upon the desired tone and emphasis. Again, the writer decides.

Introduce

The comma introduces a word, phrase, or clause of a sentence when the body of the sentence poses the question *what?* or *who?* For example:

> The president has only one secretary, Colfax.
>
> Olympic athletes have one goal, to win.
>
> The company must decide, either open a Los Angeles office or lose the California market.

Note that in each sentence the comma could be replaced by a colon or dash. Either would add emphasis to the element introduced, but the dash would add more emphasis. (**See also 5-4, p. 83, and 5-6, p. 85.**)

Note also that the meaning of the second sentence would change without a comma (or colon or dash). Instead of the athletes having a specific goal, winning, they would have to win a vague, unspecified goal. So here the comma clarifies the meaning of the sentence.

The comma also introduces letters, dialogue or quotes, and definitions, as shown below:

Dear Krista,
Jennifer said you enjoyed the Michael Caine movie...

Alfred North Whitehead has written, "The sheer statement of what things are may contain elements explanatory of why things are."

The word comma is ultimately derived from the Greek koptein, "to cut off."

Again, the colon or dash can be used in similar circumstances. In a letter a colon is more formal than a comma, a dash more informal. Use a colon in most business letters, especially to strangers; use a dash in short notes. For dialogue or quotes, the colon is used with long passages; the dash is rarely used.

In its role as an introducer, the comma also can be used to define words or concepts within the flow of a sentence. For example:

Use an ohmeter, an instrument that directly measures electrical resistance in a circuit, to check results.

The colon and dash also can be used to define. But the colon usually is used outside the flow of a sentence, in a footnote or list:

Ohmeter: an instrument that directly measures electrical resistance in a current.

The dash is usually used within the flow of a sentence, sometimes in pairs in the middle of a sentence, enclosing the definition as well as introducing it:

From a synthetic language in which meaning depended largely on inflections—internal changes in the words themselves—English has grown into an analytic language, where meaning depends largely on phrase and word order.

Note how the three concepts in this sentence are defined. A *synthetic* language is defined by the phrase "in which meaning depended largely on *inflections*."

Inflections are defined by the phrase between the dashes. And *analytic* is defined by the clause "where meaning depends largely on phrases and word order," which is introduced by a comma. (The words to be defined are underlined to emphasize that the definitions refer to them.)

5-3 DO NOT CONFUSE CONTRACTIONS WITH WORDS THAT LOOK OR SOUND ALIKE.

Contractions contain an apostrophe, a mark whose use is steadily declining. This decline increases the chance that a contraction will be confused with a look-alike or sound-alike word, or vice versa. The commonly used contractions are listed in Table 5-2.

TABLE 5-2. COMMONLY USED CONTRACTIONS

CONTRACTIONS OF PRONOUN AND VERB							
I am	I'm	I have	I've	I will	I'll	I would	I'd
we are	we're	we have	we've	we will	we'll	we would	we'd
you are	you're	you have	you've	you will	you'll	you would	you'd
he is	he's			he will	he'll	he would	he'd
she is	she's			she will	she'll	she would	she'd
it is	it's						
they are	they're	they have	they've	they will	they'll	they would	they'd
who is	who's	who have	who've	who will	who'll	who would	who'd
CONTRACTIONS OF VERB FORMS							
is not	isn't	cannot	can't	do not	don't	has not	hasn't
are not	aren't	could not	couldn't	did not	didn't	have not	haven't
was not	wasn't	will not	won't			had not	hadn't
were not	weren't	would not	wouldn't				

Like the comma, the apostrophe separates. It also distinguishes the possessive form ("a *person's* right to privacy") from the plural form ("three *persons* were hired") of a noun. **(See also 3-2.)** But the comma separates words, phrases, or clauses in sentences, and the apostrophe separates letters in words. And while the comma is used to indicate omission of a word or words in a sentence, the apostrophe indicates omission of a letter or letters in a contraction.

Some of these contractions are often confused with words that look or sound alike. The ten most common confusions are listed in Table 5-3.

TABLE 5-3. CONTRACTIONS COMMONLY CONFUSED WITH SIMILAR WORDS

it's	its
you're	your
they're	their, there
who's	whose
we're	were
we'll	well
he'll	hell
she'll	shell
can't	cant
won't	wont

The most commonly confused pair are *it's* and *its*, which both look and sound alike. *It's*, of course, is a contraction of *it is*, and *its* is a pronoun that shows possession, as in the sentence below:

> It's (It is) a luxury hotel, yet its (the hotel's) rates are reasonable.

Also troublesome are the three other sound-alikes: *you're* and *your*; *they're*, *their* and *there*; and *who's* and *whose*. *You're* is the contraction for *you are*, and *your* is a pronoun that shows possession:

> You're (You are) the boss, it's your (the boss') decision.

They're, the contraction for *they are*, is often confused with *their*, a pronoun that shows possession, or with *there*, an adverb that describes place. The differences are illustrated below:

> The evaluating team leaves for Houston tomorrow. They're (They are) going there (Houston) on their (the team's) own initiative.

Who's is the contraction for *who is,* and *whose* is another pronoun that shows possession:

> Who's (Who is) the contractor whose (the contractor's) bid was lowest?

The other pairs—*we're/were, we'll/well, he'll/hell, she'll/shell, can't/cant,* and *won't/wont*—look alike but have different meanings. *Cant,* for example, is a kind of jargon; *wont* means "accustomed." If the apostrophe is omitted, the look-alike pairs—even more than the sound-alike pairs—are likely to confuse the reader. The confusion may be momentary, but avoid even that.

5-4 WHAT FOLLOWS THE COLON EXPLAINS WHAT PRECEDES IT.

The colon introduces a list of details, or an explanation, expansion, or restatement of the statement that precedes it. Its verbal equivalent is *namely, for example, as follows,* or *that is.* If any of these fits the sentence, so does the colon:

> Technical writing has one basic function: to inform.
>
> Two important factors have not been considered: cost and date of delivery.
>
> Five inflections remain in English: -s, -ing, -ed, -er and -est.
>
> In architecture form is determined by function: form follows function.

The colon usually can be replaced by a comma or dash, the comma reducing emphasis and the dash increasing it. In example three, however, a comma would seem too much like just another comma in the series, and a dash would look awkward preceding the *-s.* And in example four a comma would be too weak because both the statement before the colon and the restatement introduced by it are complete sentences. So here the colon separates as well as introduces. A period or semicolon could be used instead of the colon (or dash), but the sense of emphatic restatement would be lost.

The colon also is used to introduce business, or formal, letters, long quotes, and definitions. (**See also 5-2, p. 74.**) And it is used to express time numerically (4:15 P.M.).

5-5 THE SEMICOLON SEPARATES; DO NOT CONFUSE IT WITH THE COLON, WHICH INTRODUCES.

The semicolon should be called a semiperiod; that would more accurately describe its single use, which is to separate. In tone and rhythm it falls halfway between the comma, which indicates a slight pause, and the period, which signals a full stop. It separates major categories in a series and independent clauses that are closely related in content, structure, or tone.

In a series that includes major categories as well as subdivisions within those categories, use semicolons to separate the categories and commas to separate the subdivisions:

> Contracts have been sent out to Hughes Airwest, PSA, and Northwest Airlines; Union Pacific, Northwest Pacific, and Southern Pacific Railroads; and Napa Valley Trucking, Pacific Motor Trucking, and Delta Lines.

The major categories, of course, are airlines, railroads, and truck lines. Commas would be technically correct between each category too. But they would blend in with the other commas. The effect would be to equalize all separations, when in fact the separation is greater between categories than within them.

Between independent clauses (that is, complete sentences), a semicolon replaces a period if the clauses are closely related:

> A United States Senate seat may cost as much as $5 million to win; the presidency may cost as much as $40 million.

Note that a period in place of the semicolon also would

be correct, but it would not emphasize the close relationship of the content of each clause.

5-6 THE HYPHEN SEPARATES LETTERS AND WORDS; THE DASH SEPARATES WORDS AND PHRASES AND INDICATES EMPHASIS OR ABRUPTNESS.

Hyphen

The main uses of the hyphen are to separate compound words (*kilowatt-hour*), to divide words that can't be completed on the same line, and to write out fractions (*two-thirds*). Since the tendency in English is to compress, the hyphen initially used in most compound words is soon dropped: *hydro-electric* becomes *hydroelectric, non-restrictive* becomes *nonrestrictive,* and so on. Triple compounds (*cause-and-effect*) usually retain their hyphens. So do compounds with a capital (*anti-American*) and most with *ex-, self-,* and *half-* (*ex-president, self-centered, half-truth*).

In dividing a word at the end of a line, the biggest problem often is where to break the word. Here are some guidelines.

1. Avoid breaking short or one-syllable words, abbreviations, contractions, acronyms (words formed from the first letters of a compound term—*N*ational *O*rganization of *W*omen), and a person's name: *omen, break, abbrev., don't, NATO, Sugarman.*
2. Divide words between syllables, using a dictionary as a guide.
3. Words like *divide,* where the dilemma is between *di-vide* and *div-ide,* break after the vowel: *di-vide.*
4. Words with double consonants (*biggest*) break between the consonants: *big-gest.* But if there is no syllable break between them (*com-pelled*), do not divide them.
5. Divide most prefixes and suffixes from the main root, or stem, of a word: *pre-cede, read-able.* **(See Tables 5-4 and 5-5.)**
6. When in doubt, check the dictionary.

TABLE 5-4.
COMMON PREFIXES: THEIR USE AND MEANING

PREFIX	MEANING	PREFIX	MEANING
ab-nor-mal	off, away	*hypo-der-mic	under
ad-vise	to, toward	in-duct	in
*ante-date	before	in-sane	not
*anti-dote	against	*inter-cede	between
bi-lin-gual	two	mis-take	wrong
com-fort	with	non-sense	not
con-flict	with	post-war	after
*contra-dict	against	pre-cede	before
de-scend	down, away	pro-ceed	forward, for
di-vide	two	re-flect	again
dis-tant	apart	sub-merge	under
dis-in-fect	not	*tele-phone	far
en-close	in	trans-plant	across
ex-port	from	tri-angle	three
*hyper-ten-sion	over	un-equal	not

NOTE: Those prefixes with * can themselves be divided on the syllable break

TABLE 5-5.
COMMON SUFFIXES: THEIR USE AND MEANING

SUFFIX	MEANING	SUFFIX	MEANING
du-ra-ble pos-si-ble	ability or likelihood	ac-tion o-pin-ion re-al-ism	act of or state of or quality of
as-sist-ance ex-ist-ence dark-ness ship-ment kin-ship	act of or state of or quality of	work-ing	act of
		home-less	without
		nerv-ous faith-ful	full of

NOTE: -ble is usually described as a variation of -able and -ible, suffixes that include the final vowel of the stem word they are attached to. But the vowel in these suffixes breaks with the stem when the word is divided.

Dash

Twice as long as the hyphen, the dash is also four times as versatile—it not only separates but ends, introduces, and encloses. Its distinctive quality in all these uses is emphasis or abruptness.

The dash is basically an alternative mark, one chosen over another for its distinctive quality. In ending, its only alternative is ellipses. Both end incomplete sentences, but the dash indicates an abrupt, slightly more dramatic ending:

> The killer waited patiently and when the victim was in range...

> The killer waited patiently and when the victim was in range—

In introducing, the dash is an alternative to the comma or the colon. (**See also 5-2, p. 74.**) It is stronger than either, and also dramatic:

> The penalty for treason in wartime is mandatory—death.

The dash here, like the colon or comma, is the equivalent of *that is* or *namely*.

In enclosing, the dash produces an emphatically different effect from its counterparts, commas and parentheses. (**See also 5-2, p. 74, and 5-7, p. 89.**) Below are three versions of the same sentence, one with commas, one with parentheses, and one with dashes:

> Three men, Einstein in physics, Freud in psychology, and Marx in economics, have shaped the modern world.

> Three men (Einstein in physics, Freud in psychology, and Marx in economics) have shaped the modern world.

> Three men—Einstein in physics, Freud in psychology, and Marx in economics—have shaped the modern world.

All three sentences are grammatically correct but different in emphasis and even in readability.

The series naming the three men and their fields is the enclosure. It is called an *appositive* series because it stands opposite the subject, *three men,* and describes it. In the first sentence the series is enclosed by the commas after *men* and *economics.* But these commas easily blend in with those after *physics* and *psychology,* and the four visually separate the sentence into five equal parts, or blocks. This misleads the reader, for the sentence consists of only two main parts—the statement that three men shaped the modern world, and the appositive series describing the men (which itself is divided into three parts). So the commas, though grammatically correct, are misleading.

In the second sentence, parentheses enclose the series. They reduce the abundance of commas and visually block the sentence into its two main parts. But parentheses play down what they enclose. The material inside them is relevant but incidental—hardly true of the names of the three men, which are the core of the sentence. So parentheses improve readability but play down an important part of the sentence.

In the third sentence dashes enclose the series. They visually block the sentence into its two main parts and play up the names of the three men, the core of the sentence. Dashes, then, are the most effective punctuation marks in this instance. Note, too, one alternative sentence structure:

> Three men have shaped the modern world—Einstein in physics, Freud in psychology, and Marx in economics.

Here the dash introduces. The colon, though less emphatic, would work as well; but the comma would still create visual problems, though not as significant.

As a separator, the dash has no full counterpart. The comma or semicolon can substitute for it in most cases, but not in all. When one of the joined elements is a fragment sentence, only the dash can separate them. Note the difference in the two sentences below:

> We want equal pay and we want it <u>now</u>—but we've been saying that for years.
>
> We want equal pay and—but what's the use?

In the first sentence a comma or semicolon could be used in place of the dash, with only a loss of emphasis. But in the second sentence only a dash will work.

5-7 USE PARENTHESES TO ENCLOSE THE WRITER'S WORDS, BRACKETS TO ENCLOSE THE EDITOR'S.

Parentheses enclose thoughts that are relevant but incidental to the sentence. They also enclose examples (as on p. 67 and 79 of this book and in footnote 1 below) that illustrate the point of the sentence. Use parentheses when to omit them and still use the material in question would alter the sense or flow of the sentence:

The future of public education (and perhaps of private education as well) depends upon the establishment of an equitable tax structure.

Brackets indicate editorial comments in a writer's manuscript.[1] Writers themselves exercise an editorial function when they quote from printed material. Those who decide to comment on, or add to, quoted material must use brackets to distinguish their remarks from the original author's. For example, a writer who uses the quote "A spokesperson for the Federal Reserve noted that Gresham's law had not been repealed" and wishes to define Gresham's law would put the definition in brackets:

"A spokesperson for the Federal Reserve noted that Gresham's law [bad money drives out good money] had not been repealed."

Brackets also are used with the word *sic,* which is Latin for "thus." When enclosed in brackets after an error, *sic* indicates that the error appeared in the original:

Acuracy [sic] is the technical writer's first responsibility and the beginning of effective writing.

Unless the material enclosed in parentheses is a complete sentence (as below), do not place any punctuation mark immediately before the second

[1]Brackets also are used to show parentheses within parentheses: (in the same year [1972]). In such cases, however, they do not represent an editorial comment.

parenthesis. A required comma, colon, semicolon, period, dash, question mark, or exclamation point goes immediately *after* the second parenthesis as in the previous sentence. (However, avoid the visually awkward combination of a dash and a parenthesis.) Any punctuation used with brackets goes inside them.

5-8 PLACE A COMMA OR PERIOD INSIDE QUOTE MARKS; A COLON OR SEMICOLON OUTSIDE.

Quote marks, originally tiny arrowheads in the margins of manuscripts, enclose dialogue and direct quotes. In the process, quote marks combine with most other punctuation marks. The comma, period, semicolon, and period are mentioned above; the others include the question mark, exclamation point, and dash. Each of these last three follows the same convention: inside the quote marks if it is part of the quote, outside them if it is not. (**See also 5-1, p. 72.**) Below are some examples of the use of quote marks with various other punctuation marks:

> "Life must be lived forward but can only be understood backward," wrote the philosopher Soren Kierkegaard.
>
> Gregory Bateson defines a paradox as "a contradiction in which you take sides—both sides."
>
> Santayana once said, "Those who cannot remember the past are condemned to repeat it"; Marx would add, "first as tragedy, then as farce."
>
> The author wrote under the section "Notes and Comments": "The names of all participants have been changed to protect their identity."
>
> "Fire Danger Area!" signs are posted throughout Sonoma County during the long, dry summer.
>
> Does anyone know who writes under the name "Adam Smith"?
>
> "A decent provision for the poor is the true test of civilization."—Samuel Johnson
>
> The personnel manager said, "When someone tells

me, 'I can do anything,' I begin to wonder if—"
"I know what you mean," the applicant said.

Note that in the last example, an exchange of dialogue, a quote within a quote takes single quote marks.

A quote may be broken up to emphasize one part of it or to create a desired rhythm. Break it, however, at a natural point:

"Love lets the other be," says psychologist R. D. Laing, "but with affection and concern."

"Love," says psychologist R. D. Laing, "lets the other be, but with affection and concern."

"If innovation is profound," educator Jonathan Kozol has written, "it is subversive."

"A lobbyist is anyone who opposes legislation I want," Senator James A. Reed of Missouri once said. "A patriot is anyone who supports me."

Note that the Laing quote can be broken in two places, as shown, depending on the desired emphasis and rhythm. The Kozol quote, however, cannot be broken elsewhere without leaving it awkward. And though the Reed quote could be broken after *lobbyist,* it would lose its parallel structure and some of its power.

EXERCISES

Read the following paragraph and—

1. Punctuate it for clarity.

The season for hurricanes is the fall for tornadoes it is the spring particularly the months of April May and June Both tornadoes and hurricanes are types of cyclones Tornadoes however unlike hurricanes are local storms last only about an hour or two and travel only about 15 miles on a path only a few hundred yards wide sometimes leveling one side of a street while leaving the other side untouched Although smaller and briefer than hurricanes

> tornadoes are more dangerous Their winds the most violent on the planet reach speeds of from 300 to 500 miles an hour and have been known to drive brittle straw through a solid wall They strike with less warning and more fury uprooting cars houses and trees indiscriminately and strewing them across the landscape

2. Excerpt from the passage the punctuated sentence, "They strike with less warning and more fury uprooting cars houses and trees indiscriminately and strewing them across the landscape." Then, where it seems most effective, break the quote with a phrase (for example, "the report states").

3. Decide whether the following three sentences require commas, and if so, how many:

> The nuclear industry which has been plagued by accidents has come under renewed public scrutiny.
>
> The president who was one of the founders of the company retired last week.
>
> Those auto manufacturers that offered cash rebates to new car buyers raised their sales volume substantially.

4. Punctuate the following sentence first to emphasize the three activities, then to underplay them:

> Those activities that contribute to an effective report planning research and revision are overlooked by many writers.

5. Punctuate the following sentence:

> The following cities are being considered as the site of the Society's annual convention Rochester New York Charleston South Carolina Olympia Washington and Council Bluffs Iowa.

Despite the emphasis on format in the teaching of technical writing, the meaning of the word remains blurred. Sometimes it refers to the physical appearance of a report, sometimes to the relative formality of a report, sometimes to the type of report—proposal, progress, laboratory, and so on.

The word itself is derived from the Latin *formatus*, which means "formed." Originally, the phrase *liber formatus* described a book formed in a certain shape and size. *Format* is still used in publishing to refer to shape and size (and today to layout too). But the broader sense of the word, as defined by *Webster's Third*, is "a general plan or arrangement of something," and that is the meaning followed here.

The specific format of a piece of technical writing is shaped by the purpose of the writer and by the piece's audience. Generally, however, most technical writing is structured along the traditional lines of (1) an introduction, (2) the body of the piece, and (3) a conclusion. The introduction orients the reader; it includes any necessary background and describes the intent and scope of the piece. The body contains the *what* and *how* and *why* of the subject. And the conclusion ties everything together: the writer summarizes and, if necessary, evaluates and recommends.

There are no universal rules of format, but one basic precaution is worth remembering: format should serve the writer, not vice versa.

6-1 TITLES SHOULD BE PRECISE AND DESCRIPTIVE, EVEN LONG IF NECESSARY, BUT NOT CUTE OR CLEVER.

The main function of a title is to describe instantly for the reader the subject of a report or paper. In doing so it plays a curiously dual role. As the first words the reader sees, the title is like a condensed introduction: it sets the direction, the scope, even the tone, of what follows. The same title, however, is also the most concise summary of what follows.

To function effectively in these roles, a title must be inclusive without bogging down in detail, as in the examples below:

FATIGUE IN ROCK WEATHERING

CONTROLLING THE CARAGANA APHID

Turbidity, Currents, and Sediments
in the North Atlantic

Communication Methods
Used by Hearing-Disabled Students
at Santa Rosa Junior College

A Psychoanalytic View of Hostility:
Its Genesis, Treatment,
and Implications for Society

Note that titles may be written in capital letters (either closely or widely spaced) or in capitals and lower-case letters. When using upper- and lower-case letters, capitalize only nouns, pronouns, verbs, and modifiers, as shown above. Do *not* capitalize conjunctions (*and, or,* for example), articles (*the, a,* for example), or prepositions (*of, in, for,* and so on) unless they are the first word of the title or they contain more than five letters.

Also note that if the titles for examples one and two had simply been ROCK WEATHERING and THE CARAGANA APHID, they would have been too general to describe accurately the scope of the report. These titles were easily made more specific. Another way of making a title more specific is shown in example five. There the colon allows expansion of the title to include a description of the scope of the psychoanalytic view of hostility.

Finally, since titles succinctly summarize what has been written, it is wise to choose one *after* completing the writing, so as to reflect the final shape of the report or paper.

6-2 FORMAL REPORTS ARE LONG, INCLUSIVE, AND ELABORATE; INFORMAL REPORTS ARE LESS SO.

The difference between a formal and an informal report is one of degree; there is no hard dividing line. A formal report is tightly structured and usually—but not necessarily—10 pages or longer. It often serves as a permanent record, and it usually is written for a relatively wide audience. While a formal report follows the general report format of introduction, body, and conclusion, it also includes as many as 11 features infrequently found in an informal report. These features are listed in Table 6-1 in their usual sequence.

TABLE 6-1. FORMAT OF FORMAL AND INFORMAL REPORTS COMPARED

FORMAL REPORT FEATURES	USE IN INFORMAL REPORTS
(1) Cover	Very rarely
(2) Separate Title Page	Rarely
(3) Letter of Transmittal	Rarely
(4) Foreword or Preface	Very rarely
(5) Table of Contents	Rarely
(6) List of Illustrations	Rarely
(7) Abstract or Summary	Sometimes
(8) Appendix	Rarely
(9) Glossary	Rarely
(10) Bibliography	Sometimes
(11) Index	Very rarely

1. The *cover* of a formal report usually includes the title of the report, the company's name, the date, and a file number. It also may include the author's name.

2. The *title page* repeats the title of the report, and the date and file number, and adds the author's name. It may also include an *approvals* or *distribution* list—names of people who must approve the report or receive a copy of it. These lists also may be placed on a separate sheet following the title page.

3. The *letter of transmittal* is a kind of introduction to the report. It refers an individual to the report; briefly explains its relevance, purpose, and scope; acknowledges vital assistance; and closes with the hope that the report will be satisfactory—and all in one page. The letter usually is separate from the report, but it may be bound into it. If the report goes to several readers with different interests, however, send a separate letter to each reader, tailoring each letter to the interest of the particular reader.

4. The *foreword* or *preface* also serves as an introduction. Its function is similar to that of a letter of transmittal, but a foreword or preface is less personal. Either or both may be used, with or without a letter of transmittal. The choice depends on company policy, on the nature and purpose of the report, and on the known preferences of the audience.

5. The *table of contents* outlines the content, scope, and organization of the report and anything else the writer decides is helpful to the reader.

6. The *list of illustrations* identifies and locates all graphics—charts, tables, graphs, drawings, diagrams, figures, and photos. Illustrations also are used in informal reports, but they are rarely listed separately.

7. The *abstract* briefly explains what the report is about; the *summary* condenses what the report says. The distinction, however, is often blurred, and either term may be used to refer to a brief description of the contents of a report. Either is placed *before* the body of the report. (**See also 6-5, p. 110.**)

8. The *appendix* follows the body of the report. It contains any material that is relevant to the report but does not fit into its main sections. The appendix (or *appendices*) may include charts, tables, samples, documents, equations, or references.

9. The *glossary* is a select dictionary of technical terms used in the report. It is arranged alphabetically and sometimes listed as an appendix.

10. The *bibliography* lists the sources—the books and publications—used in researching the report or paper.

Like the glossary, it is arranged alphabetically and may be listed as an appendix. *Footnotes* also may be gathered together and added to the bibliography or listed as a separate section or as an appendix. (**See also 6-8, p. 124.**)

11. Only book-length reports or papers contain an *index*, an alphabetical list of names and subjects with the page number where they are discussed. Such reports, however, are not uncommon, especially in government.

A formal report contains some or all of the above features; an informal report may contain some of them. *Why* a report is formal or informal, however, is harder to pinpoint. Length alone is not necessarily decisive; some informal reports are 30 pages long, some formal ones only 8 pages. The desire to impress and the nature of the report itself are important factors. Formal reports generally are broader in scope than informal reports and usually are written for management or for an audience outside the company—customers, shareholders, or the public, for example. Informal reports usually are written for an audience within a department —either the writer's or the reader's. They also usually focus on procedure, while formal reports lean toward results. In short, the factors that determine whether a report is formal or informal reflect—as they should in technical writing—the purpose of the report and the nature of the audience.

6-3 THE USUAL FORM OF AN INFORMAL REPORT IS A LETTER OR MEMO; THE MEMO IS THE MORE INFORMAL OF THE TWO.

Most technical writing is informal. The informal report ranges from a time sheet to a detailed description of an industrial process. Between these extremes lie sales and personnel reports; surveys and proposals; inquiries and complaints; accident, progress, and research reports— and more. For most of these, a one- or two-page letter or memo is the standard format.

Letters

Although a letter is more formal than a memo, it is still informal—and personal. It is like a conversation between the writer and reader in which the writer does all the talking. To know what to say and how to say it, then, *the writer must understand the reader's point of view*. That is the most important element of letter writing—and the most difficult to grasp.

Letters—even business letters—are more personal than other written communications. That makes them potentially more powerful—and more risky. Powerful if the writer touches those qualities in the reader that inspire good will and confidence. Risky because the writer may strike a sore spot in the reader that triggers fear or anger. Indeed, the reader may be affected more by a letter's tone than by its content. For example, whether a letter is sincere or affected, assertive or aggressive, may make it or break it. Even a near-miss here risks losing the reader's good will—and business.

This risk is one reason why the use of personal pronouns like *I* and *you* traditionally has been discouraged in business and technical letters. To eliminate this risk in business correspondence, the personal element was removed from it and replaced by the impersonal passive voice. A reader could not be offended by someone who wasn't there.

Most business and technical letters inform, but many persuade, and some instruct. Still others combine these functions. Whatever its function, however, a letter can be divided into the three basic parts of any piece of technical writing: (1) an introduction, (2) the body, and (3) a conclusion. The special application of these categories to letters is examined below. Figure 6-1 shows a common format for typed letters. Note that the letter is balanced and spaced evenly; has roughly equal, one-inch margins; and that the complimentary close lines up flush with the writer's address. In the semiblock format shown, each paragraph usually is indented five spaces. In the block format, a variation of the semiblock format, the writer's address and the complimentary close are moved to the left-hand margin, flush with the inside address; nothing is indented. This saves typing time by eliminating use of the tab key, a saving that can be substantial if correspondence is heavy.

FIGURE 6-1.
A SAMPLE LETTER FORMAT

```
                                    1978 McGeorge Place
                                    Berkeley, CA 95400
                                    April 28, 19--

(Doublespace twice)
    Mr. Richard DeLouise
    Senior Staff Economist
    The Corporation for Economic Research
    109 Cherry Street
    Washington, D.C. 20025
(Doublespace)
    Dear Mr. DeLouise:
(Doublespace)
         The introductory paragraph.  It should establish
    contact with the reader.
(Doublespace)
         The body of the letter.  Doublespace between
    paragraphs, single space within them, unless the
    letter is very short.
(Doublespace)
         The concluding paragraph.  It should leave a
    legacy of good will with the reader.
(Doublespace)
                                    Sincerely,

(Doublespace twice)

                                    Margaret Hoehn
```

1. The *introduction* of a letter includes the inside address, the salutation, and an opening paragraph. The inside address consists of the addressee's name and title (if any) and the company's name and address:

> Mr. Richard DeLouise
> Senior Economic Policy Analyst
> The Government Research Corporation
> 1730 M Street, N.W.
> Washington, D.C. 20036

In the salutation use *Dear Mr.* or *Dear Ms.* followed by the person's last name and a colon (*Dear Mr. DeLouise:*). Avoid *Mrs.* or *Miss*—a woman's marital status is as irrelevant as a man's. When unsure how a woman may react to *Ms.*, or where a first name may belong either to a man or woman, use the person's full name (*Dear Robin Morgan*) followed by a colon.

The traditional salutations used in writing to a company blindly are *Gentlemen* or *Dear Sir* (or *Sirs*), followed by a colon. But these salutations exclude women and increasingly are considered sexist. One alternative is to direct the letter to a specific department, using the title of the department head (*To the Sales Manager*) or the salutation *To Whom It May Concern*, followed by a colon. Another alternative is to obtain the name of the department head, perhaps by calling the company and asking the switchboard operator. If the letter is important, the money a call costs will be well spent.

A letter report often has, in capitals, a subtitle (SUBJECT) to indicate the subject of the report and subheads (REQUIREMENTS, CONCLUSIONS, and so on) to introduce its sections. Place the subtitle between the inside address and the salutation; place the subheads in the body of the report. Both should be flush with the left-hand margin.

Use the opening paragraph of a letter to establish contact with the reader. Remind the reader of any previous personal contact—a meeting, a telephone conversation, correspondence. Mention any mutual friend or contact (but first get their permission). Refer to previous meetings, conversations, or correspondence by date; acknowledge receipt of a letter with thanks. For example:

You may recall that we met in Washington last April at the...

I enjoyed talking to you on the telephone Wednesday. As you suggested, I have...

Thank you for your letter of October 2.

Ray Reinhardt suggested that I write to you.

Your associate, Patrick Murphy, may have mentioned my name to you.

When contact has been established with the reader, explain the purpose of the letter.

If there has been no previous personal contact, and

there are no mutual friends, simply explain the purpose of the letter.

2. The *body* of a letter contains the message: what the letter is about. Most letters fall into one of three categories: (a) inquiries and the response to them; (b) complaints and the adjustment of them; and (c) proposals and sales letters. The first two involve information and sometimes instruction; the third involves information and persuasion.

Whether writing an inquiry or a complaint, or a response to either, be specific and direct. Use personal pronouns—even *I*—when to avoid them would create an artificial tone or awkward sentence. Keep paragraphs short; cover the subject thoroughly yet concisely. Be tactful where necessary, reserved in response to bluntness or sarcasm. If you have made an error, admit it—no one is perfect—and offer to adjust it. Most important, keep the reader's point of view in mind.

The above also applies to proposals and sales letters. But here persuasion is involved, so the first question to answer is: *What do you want the reader to do*? Then cite reasons and evidence that may persuade the reader to do it. Try to anticipate serious objections and address them. Emphasize the reader's prospective benefits, but do not promise more than you can deliver.

3. The conclusion of a letter includes a final paragraph and a complimentary close.

Use the final paragraph to close on a positive note. It may be as simple as a personal expression of thanks or hope or best wishes for success. The final paragraph also may briefly summarize the contents of the body of the letter or repeat an important point. And it may suggest further review of the matter, possible action, further communication, or a personal meeting.

The complimentary close should reflect the degree of informality or intimacy between the writer and reader. It includes the words *truly, respectfully, cordially,* or *sincerely,* usually combined with *yours,* sometimes with *very,* and followed by a comma. The most frequently used combinations are *Yours truly, Sincerely yours,* and simply *Sincerely.*

Figures 6-2, 6-3, and 6-4 on p. 104, show examples of a letter of inquiry, a letter of complaint, and a letter of

FIGURE 6-2. A SAMPLE LETTER OF INQUIRY

1440 Slater Street
Santa Rosa, CA 95404
January 5, 1979

Mr. Howard Greenwald
Director of Communications
Data Processing Division
IBM Corporation
White Plains, New York 10604

Dear Mr. Greenwald:

You may recall that a few years ago I wrote for Data Processor an article on IBM's "on-line" customer inquiry service for the Brooklyn Union Gas Company.

Now a teacher of technical writing, I am completing a textbook, The Elements of Technical Writing, for Harcourt Brace Jovanovich. I am writing to you for some samples of technical writing for use in the text.

Specifically, I have two requests. First, permission to quote briefly from my own article in Data Processor. Second, three or four samples of technical reports from IBM files, unclassified of course. I am especially interested in reports of two to five pages, either descriptive or instructional. I would need them by February 15, 1979.

Of course, IBM will be fully credited for any material used in the book, and I will check with you for approval of any minor changes we may wish to make -- for example, in length.

Any assistance you can give me will be much appreciated.

Sincerely,

Joseph Alvarez

Here the first paragraph is used to remind the reader of a previous association; the second paragraph, to explain the purpose of the letter; the third paragraph, to describe specific requests; and the fourth paragraph, to close on a polite note of appreciation for whatever help might be available. (NOTE: Mr. Greenwald sent the material I asked for, but I decided not to use it.)

adjustment replying to the complaint. Note that in each case the writer requires only a page, the usual length of most inquiry, complaint, and adjustment letters.

FIGURE 6-3. A SAMPLE COMPLAINT LETTER

P. O. Box 84
Rohnert Park, CA 94928
November 14, 19--

Consumer Service Department
Furniture Inc.
96 Powell Street
San Francisco, CA 94108

Dear Sirs:

On October 6, 19--, Furniture Inc. delivered a brass and glass coffee table I had purchased in March, order #SF24950. The table arrived with the glass severely scratched, and I immediately wrote to you requesting that the glass be replaced or the table exchanged.

I have had no response from your office, though the problem was brought to your attention more than nine weeks ago.

I still wish to have the glass replaced or the table exchanged, and I urge your office to take care of this matter within two weeks.

Sincerely,

Nadine Bruns

In this letter, the writer includes the purchase order number to help the company locate its records. By using the word immediately instead of the exact date of her first letter, she emphasizes the promptness of her request—and also subtly underlines the contrast between her promptness and the company's inaction. Finally, she repeats her request and politely but firmly makes it clear she expects action within two weeks.

FIGURE 6-4. A SAMPLE LETTER OF ADJUSTMENT

FURNITURE INCORPORATED
96 Powell Street
San Francisco, CA 94108
November 25, 19--

Nadine Bruns
P.O. Box 84
Rohnert Park, CA 94928

Dear Nadine Bruns:

Please accept our apology for the delay in replying to your letter of over a month ago.

I have again contacted the company which manufactures the coffee table you purchased from us. A duplicate of your table will be shipped to your home the week of December 11, 19--, via our company carrier. At that time we will pick up the damaged coffee table you now have.

I hope you will enjoy your new coffee table for many years, and that you will consider Furniture Inc. for any future purchases of fine home furnishings.

Sincerely yours,

Jean Corvin
Consumer Service Department

The writer addresses this letter to "Nadine Bruns," since that is the way she signed herself, and opens with an apology. In the next paragraph, by using the word again, the writer implies that the delay was the manufacturer's fault. Finally, she closes on a friendly note in an attempt to restore Nadine Bruns' good will toward Furniture Inc.

Finally, it seems appropriate here to examine a special kind of letter, one that most of us write at one time or another: a job application letter. One of the most difficult letters to write well, it must both inform and persuade, and the writer must steer a narrow course between sounding boastful and seeming ordinary.

The guidelines for all letters apply here: (1) establish contact with the reader; (2) be specific and direct, thorough yet concise; (3) use personal pronouns; and (4) keep paragraphs short. For a job application letter, also, (5) cite reasons and evidence why you qualify for the job; (6) emphasize the benefits to the employer of hiring you; and (7) don't promise more than you can deliver.

Before you write, however, *you must ask yourself who you are and what you want.* Specifically, what are your special qualities? Your strengths and weaknesses? Your limits? Are you a leader or follower, outgoing or introverted? What are your goals? Wealth? Excitement? Prestige? Glamour? Freedom? Challenging work? Security? Fame? Power? Why do you want *this* job?

The employer need not know the answers to these questions—though a keen interviewer will certainly probe for some of them—but *you* should know them. Knowing who you are and what you want helps you deal with the world and the people in it more effectively. You must of course be honest with yourself, even if the truth is hard to accept. Ignorance of self leaves one vulnerable from without; but a false image of self subtly undermines one from within.

A job application letter not only must inform and persuade, it also must compete with many others for the attention and interest of the reader. This challenge is most effectively met by separating the information from the persuasion. Place all personal information (education, work history, and so on) on a separate *data sheet* (sometimes called a *vita* or *resumé*). With this data sheet, send a covering letter explaining your interest in the job and your qualifications for it—the decisive elements of an application. Isolated, these elements hold the full attention of the reader, who can refer to the data sheet if necessary to measure your qualifications.

The data sheet should be representative, not exhaustive. Prepare it specifically for each prospective job; don't use an all-purpose resumé. List important experience, schooling, personal data, and any accomplishments relevant to the job. References are optional; they may not be required. In your covering letter you can offer to supply them if the reader wants them.

When preparing more than one data sheet, photocopy as many as you need, preferably on beige or gray paper (which will stand out in a pile of white sheets). But type every covering letter individually.

Personal data includes your full name and address and telephone number. Age, health and marital status, and dates and length of military service (if any) are optional, so include what will help you and omit what won't. For example, an employer may equate marriage with stability and responsibility in a man but consider it a liability in a woman, thinking she would probably resign if she became pregnant or her husband had to relocate.

List your education in reverse chronological order, high school last. The name of each school, the year of graduation, and your college degree and major are enough. Include military schooling if it is relevant and technical training if it is relevant and the school is accredited. If your rank in your graduating class is impressive, mention it, especially where it is important, as in the legal and medical professions. If you hold a degree beyond the B.A. or B.S., you need not list your high school.

List work experience in reverse order—the most recent job first. Include the beginning and ending dates for each job, the employer's name, your title (if any), and a brief description of your duties (unless your title describes them). Do not list insignificant jobs. If you don't have much work experience, expand the section on education; for example, list relevant technical classes, workshop projects, and so on. Figure 6-5 shows one format for a data sheet.

FIGURE 6-5. A SAMPLE DATA SHEET (RESUMÉ)

James Muckle
2800 Ocean Way
Santa Barbara, CA 93101
(805) 216-1950

PERSONAL DATA

Age: 28
Health: Excellent

EDUCATION

1978 Nevada Teaching Certificate
1978 California Teaching Credential (Ryan Single Subject: English)
1976 Sonoma State University, Rohnert Park, CA, B.A. (English)
1973 Santa Rosa Junior College, A.A.

TEACHING EXPERIENCE

September 1978 - June 1979: English teacher, Bonanza High School, Las Vegas, Nevada.
June 1978: Sonoma County Coordinator for California Lutheran College Lecture Series for Educators. Organized and moderated the summer session series.
February 1978 - June 1978: Substitute teacher for Santa Rosa City Schools and Petaluma City Schools, both in California. Also a Home and Hospital teacher for the Santa Rosa District.
September 1977 - February 1978: Student English teacher at Rancho Cotate High School, Petaluma, CA.

WORK EXPERIENCE

1971 - 1978: Various part time jobs, including office manager, copy machine repairman, short order cook, bartender, and sales clerk.
1969 - 1971: Merchant seaman. Shipped out to the Far East, South and Central American, Alaska, Mexico, and points between.

The covering letter that is sent with the data sheet should be typed and should not exceed one page. Use the opening paragraph to establish contact with the reader. Show an interest in the job and, if possible, a knowledge of the company. Such knowledge, rare in job application letters, will immediately mark you as special in the eyes of the reader.

In the body of the letter state your qualifications for the job. Emphasize your strength. If it is experience, select the one or two most impressive, most relevant items from your resumé ("As you will note in my resumé, I have . . .") and explain how they will apply to the job. Stress any related experience and achievements in the field. If you lack impressive experience, focus on academic accomplishments. Or if you have the personal traits employers look for—enthusiasm, initiative, eagerness to learn—emphasize them. Remember, employers hire for their own reasons, not for the applicant's, so keep the reader's point of view in mind. Do not discuss salary, fringe benefits, vacations, or working conditions; save those for an interview.

Use the final paragraph to set up an interview at a mutually convenient time. Also mention when you will be available for work, and offer to supply references and any further information the employer may require. If the application should be kept confidential for any reason, the final paragraph is the place to request that.

Memos

The most informal of formats—the memo—is immediate, concise, almost telegraphic. It may run from two lines to two pages, but rarely more. Where letters are used to correspond outside a company, memos are used within a department or between departments of the same company. So a memo need not, like a letter, establish contact with the reader; the contact is already there. Nor does a memo contain a complimentary close.

Below is a typical memo format, part of which is usually preprinted in the upper-left-hand corner of the page, either flush with the left-hand margin or with the colons aligned, as in the example below. (The abbreviation *cc* stands for "carbon copy.")

TO: Matt Milan
FROM: Joe Alvarez
cc: Marlowe Teig
Jim O' Donnell
Tom Frazier
DATE: January 25, 19—
SUBJECT: TESTING OF NEW COIN-SLOT POOL TABLES

Use a memo where the subject is minor or is a single aspect of a large project, or when the subject is of limited scope or of passing importance. Without preliminaries, state the message of the memo; develop it as much as necessary; and, if applicable, add any recommendations or conclusions. As in a letter, be specific and direct, thorough, yet concise. And use subtitles (in capitals) to introduce important sections.

6-4 DO NOT DISCUSS THEORY OR TRANSCRIBE EVERY DETAIL OF A LAB NOTEBOOK IN A LAB REPORT.

The lab notebook is the *complete,* chronological record of everything that happens in the lab. It includes the following: (1) a brief explanation of the nature and purpose of the lab work; (2) a list of the technicians involved; (3) a detailed description of the equipment used; and (4) any test procedures (what, where, and how) and the test results. The lab *report* contains *selected* information from the lab notebook, the selection depending upon the report's purpose and audience.

A lab report may be formal or informal, depending upon its complexity and length. Since it is essentially the description of a process, organize it like a lab notebook: nature and purpose of the process; technicians involved; apparatus used; test procedures; and test results. But report selectively and write concisely.

Record the report number, the date, and your name at the beginning of a lab report; place special observations, analyses, or conclusions at the end.

6-5 AN ABSTRACT EXPLAINS WHAT A REPORT IS ABOUT; A SUMMARY CONDENSES WHAT A REPORT SAYS.

The table of contents and the abstract are the two most widely read sections of a formal report. The word *abstract* itself, however, is loosely used by technical writers: sometimes it is, and sometimes it is not, a summary. And even when it is a summary it still may be called an abstract. Confusing.

The key to the distinction between an abstract and summary lies in the function of abstracts. A *topical,* or *descriptive,* abstract explains what the report is about; it does not include results, conclusions, or recommendations. An *informative* abstract condenses what a report says, including its results, conclusions, or recommendations. The informative abstract sometimes is called a *summary,* thus the confusion.

In short, an abstract explains selectively, a summary condenses inclusively. Another perspective: an abstract is a short summary, a summary a long abstract. Either or both are used in a formal report; one or the other is sometimes used in an informal report; and either precedes the actual report or technical paper.

When writing an abstract or summary, focus on the main idea of the report and cover all important points (adding to a summary any results, conclusions, or recommendations). Use abbreviations freely and numerals (even under 10) but no tables, graphs, or charts. Limit an abstract or summary to one page, unless the report is very long. In that case hold the abstract to 5 to 10 percent of the length of the report, the summary to 10 to 20 percent of the length. And write them after the report or paper is completed.

Figures 6-6 and 6-7 show an abstract and a summary of Section 6, "Format," of this book. Note the difference between the two in approach and length.

FIGURE 6-6. AN ABSTRACT OF SECTION 6, "FORMAT," OF THE ELEMENTS OF TECHNICAL WRITING

Format is defined, and the general format of a report is described. The function of titles is explained. Formal and informal reports are compared in terms of length, scope, audience, and special features.

The personal nature of letters and its possible effect on the reader is examined. Letters are analyzed in terms of tone, content, and format. Inquiry, complaint, and adjustment letters, and job application letters and resumés, are examined and illustrated.

The uses and format of the memo are described. Lab reports are compared to lab notebooks in format and content. Abstracts and summaries are defined, compared, and illustrated.

The use of tables, graphs, charts, drawings, and diagrams are described and illustrated, with emphasis on the importance of accuracy, relevance, and labeling of illustrations. When to use numerals and when to write out numbers is explained, and consistency in the use of numbers is illustrated. The format and uses of footnotes and bibliographies are described and illustrated.

FIGURE 6-7. A SUMMARY OF SECTION 6, "FORMAT," OF *THE ELEMENTS OF TECHNICAL WRITING*

Format here means "a general plan or arrangement of something," and it should serve the writer, not vice versa. The general format of reports is: introduction, body, conclusion. A title, which functions as a condensed introduction, should be precise and descriptive.

The difference between a formal and informal report is in degree. A formal report usually is longer, more permanent, and broader in scope. It also contains features, such as a table of contents, seldom found in informal reports.

A letter is the most personal of informal reports, which makes it potentially more powerful— and more risky. Letter format includes an introduction, body, and conclusion: the introduction to establish contact with the reader, the body for the message, the conclusion to close on a positive note. Inquiry, complaint, and adjustment letters should be specific, direct, and brief. Job application letters must inform and persuade, also compete with others.

A memo, the most informal of reports, is immediate, concise, almost telegraphic. A lab report contains selected information from the lab notebook, not theory. An abstract explains what a report is about; a summary condenses the contents of a report.

Tables present data in a convenient format; graphs and charts show trends or emphasize comparisons; drawings and diagrams help clarify descriptions. Illustrations should be accurate, relevant, and clearly labeled. The standards for when to write out numbers or use numerals vary, but they should be consistent within each piece of writing. In a report, footnotes comment on or further explain the text; a bibliography lists reference sources.

6-6 USE TABLES TO PRESENT RELATED DATA, GRAPHS TO SHOW TRENDS, AND DRAWINGS TO CLARIFY DESCRIPTIONS.

Illustrations are sometimes necessary and almost always helpful. They clarify descriptions, emphasize comparisons and contrasts, project trends, and outline relationships. They also add variety and change of pace to a report or paper, making it more interesting and easier to read. In large companies, illustrations usually are provided by a drafting department or by technical illustrators. The writer, however, is responsible for the illustration's accuracy, relevance, and effect.

When using illustrations, make sure they are accurate and relevant as well as simple enough to understand. Use a scale neither too large nor too small, one that clarifies the trend or comparison rather than exaggerates or obscures it. Number and caption each illustration, and acknowledge the source of any that are not original. Finally, insert them into the text, unless they interrupt its flow.

Among the many types of illustrations available to technical writers, the three most commonly used are: (1) tables; (2) graphs and charts; and (3) drawings and diagrams. In the text, tables are captioned *Table* and numbered; all other illustrations are captioned *Figure* and numbered. The three basic types are discussed briefly below:

Tables

Tables present data or statistics in a convenient, precise format that makes the data easier to read and understand, as in Table 6-2. Check all data for accuracy, and space it so it does not look crowded or jumbled. Use consistent units of measure throughout, and caption all columns. If additional information is needed to clarify some aspect of the data, place it beneath the table, prefaced by the word *NOTE* (in capitals) and a colon.

Tables are often used in conjunction with graphs; in fact, many graphs are plotted from the data in tables. In such instances the graph dramatically shows a trend documented by the data in the table.

TABLE 6-2. PROJECTED REQUIREMENTS AND JOB OPENINGS

OCCUPATIONAL GROUP	1974 EMPLOYMENT	PROJECTED 1985 REQUIREMENTS	PERCENT CHANGE	OPENINGS, 1974—85		
				Total	Growth	Replacements
TOTAL	**85,936**	**103,400**	**20.3**	**57,600**	**17,400**	**40,200**
White-collar workers	41,739	53,200	27.5	34,300	11,500	22,800
Professional and technical workers	12,338	16,000	29.4	9,400	3,600	5,700
Managers and administrators	8,941	10,900	21.6	5,200	1,900	3,200
Salesworkers	5,417	6,300	15.7	3,400	900	2,600
Clerical workers	15,043	20,100	33.8	16,300	5,100	11,300
Blue-collar workers	29,776	33,700	13.2	12,500	3,900	8,600
Craft and kindred workers	11,477	13,800	19.9	5,100	2,300	2,800
Operatives	13,919	15,200	9.0	6,000	1,300	4,800
Nonfarm laborers	4,380	4,800	8.8	1,400	400	1,100
Service workers	11,373	14,600	28.0	11,000	3,200	7,800
Private household workers	1,228	900	-26.7	600	-300	900
Other service workers	10,145	13,700	34.7	10,400	3,500	6,900
Farm workers	3,048	1,900	-39.0	-100	-1,200	1,000

NOTE: Numbers in thousands. Detail may not add to totals because of rounding.

SOURCE: Occupational Projections and Training Data (Washington, D.C.: U.S. Department of Labor, Bureau of Labor Statistics, 1976) Bulletin 1918, p. 9.

Graphs and Charts

Graphs, sometimes called charts, show trends or emphasize comparisons or contrasts. They vary in form and function. A line graph shows trends; a bar graph emphasizes comparisons or contrasts; a circle graph dramatizes the distribution of something. Flow charts indicate movement or direction in a process or system, and organizational charts outline responsibilities and relationships in a company or system.

Graphs are often plotted from the statistics contained in tables, and sometimes shown with the tables. As with tables, check graphs for accuracy, and place additional information in a note beneath the drawing. Also, clearly label important features of a graph, on the drawing itself when necessary.

FIGURE 6-8.
UNITED STATES YEARLY INFLATION RATE, 1960-1978

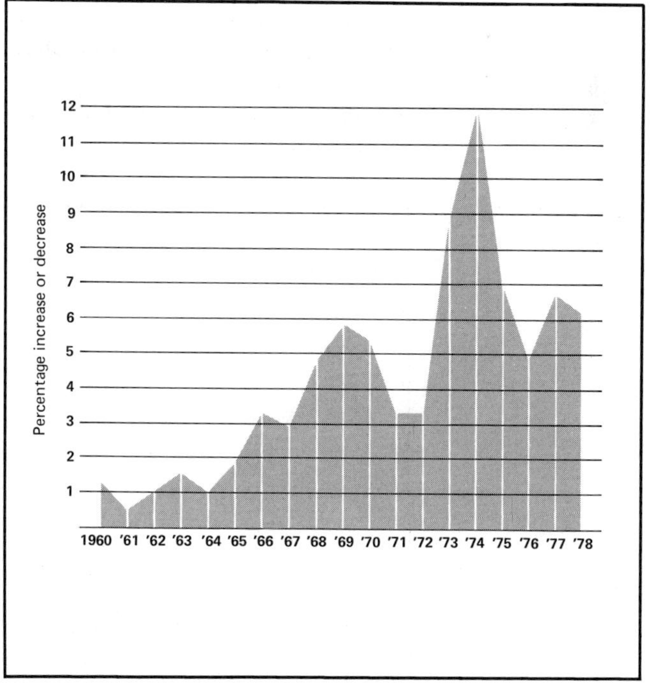

NOTE: The scale at the left indicates the percentage increase or decrease; figures for 1978 include the first 8 months.

SOURCE: U.S. Bureau of Labor Statistics

Line Graphs. A line graph contains two variables—that is, factors that change according to the interaction between them. The *independent* variable usually is plotted on the horizontal, or X, axis, the *dependent* variable on the vertical, or Y, axis. As the independent variable changes, it affects changes in the dependent variable. In Figure 6-8, the independent variable is time; the dependent variable is the rate of increase of inflation. The rate of increase of inflation changes with the passage of time.

Once the changes are plotted and connected by lines, the trend produced by the interaction of the two variables becomes clear. This trend can even be projected into the future—without, however, any guarantee of accuracy, as countless economists have discovered.

The trend of various consumer items also could be traced on Figure 6-8, using dotted lines, dashes, and even colored lines to distinguish them. When more than one line appears on a line graph, however, they all must be labeled; and more than four such lines invites confusion.

FIGURE 6-9.
RAINFALL IN SONOMA COUNTY, 1975-1978

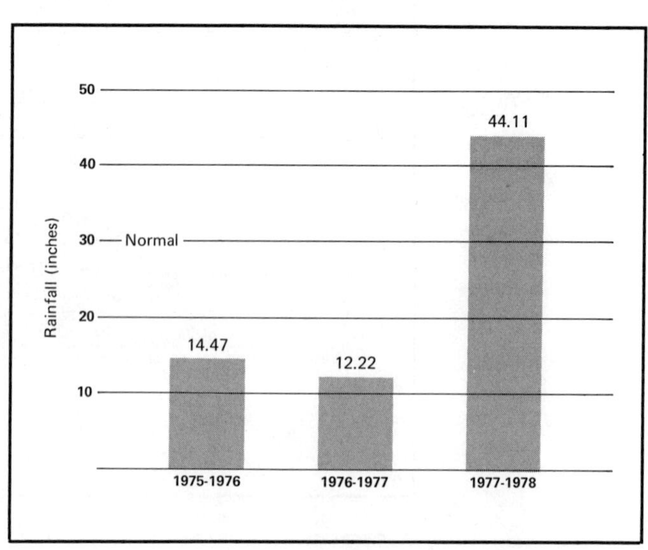

SOURCE: Rainfall figures from Sonoma County (Calif.) Water Department.

Bar Graphs and Circle Graphs. A bar graph effectively emphasizes differences in values or amounts between or among several items. Figure 6-9 vividly illustrates the difference in rainfall in Sonoma County between the drought years of 1975-1977 and the 1977-78 season. Note the spaces between the bar; they add clarity. The figures also are placed on the bars themselves, standard procedure with bar graphs.

Circle graphs are sometimes called pie graphs because they resemble a sliced pie. In Figure 6-10 the pie is the annual federal budget, and the graph shows how federal funds are collected and distributed. Note that the segments of the two graphs are placed clockwise around the circle in order of size, the largest first, just to the right of the noon position. This is the usual order of data in circle graphs.

FIGURE 6-10. THE ANNUAL FEDERAL BUDGET, 1970-1977

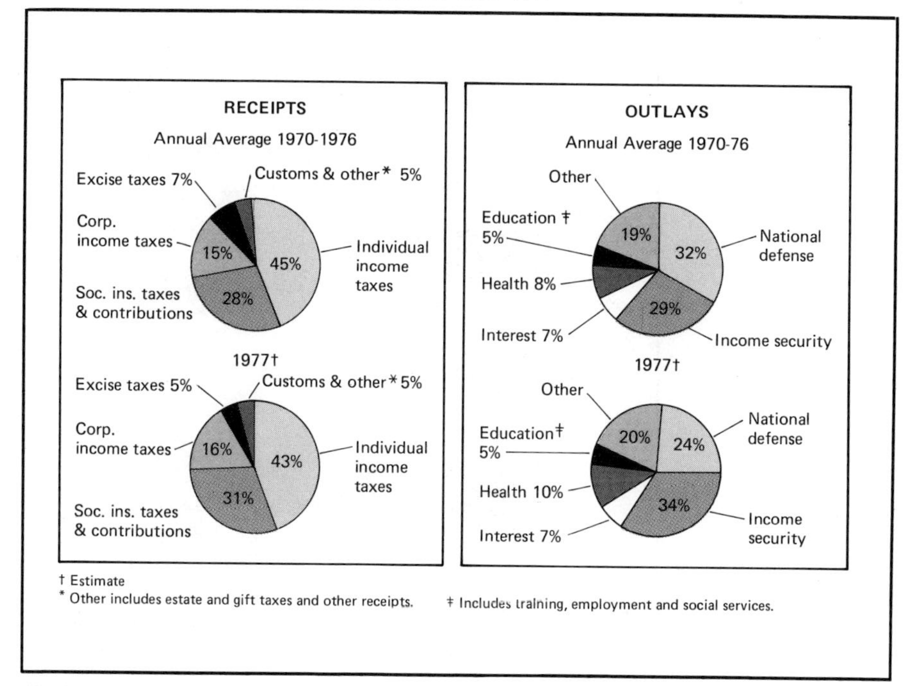

SOURCE: Statistical Abstract of the United States (Washington, D.C.: U.S. Bureau of the Census, 1978), p. 240.

Flow Charts and Organizational Charts. Flow charts contain tiny arrows that indicate the direction of flow through the various stages of a process or system. The simplest flow charts are drawn with squares, circles, or rectangles, as in Figure 6-11. In electronics, simplified flow charts of circuits or modules are called *block diagrams;* the completely annotated versions are called *schematic diagrams* (see Figure 6-12).

An organizational chart (see Figure 6-13) outlines the administrative structure in a company or system. Though it does not show movement, it indicates the flow of administrative responsibility.

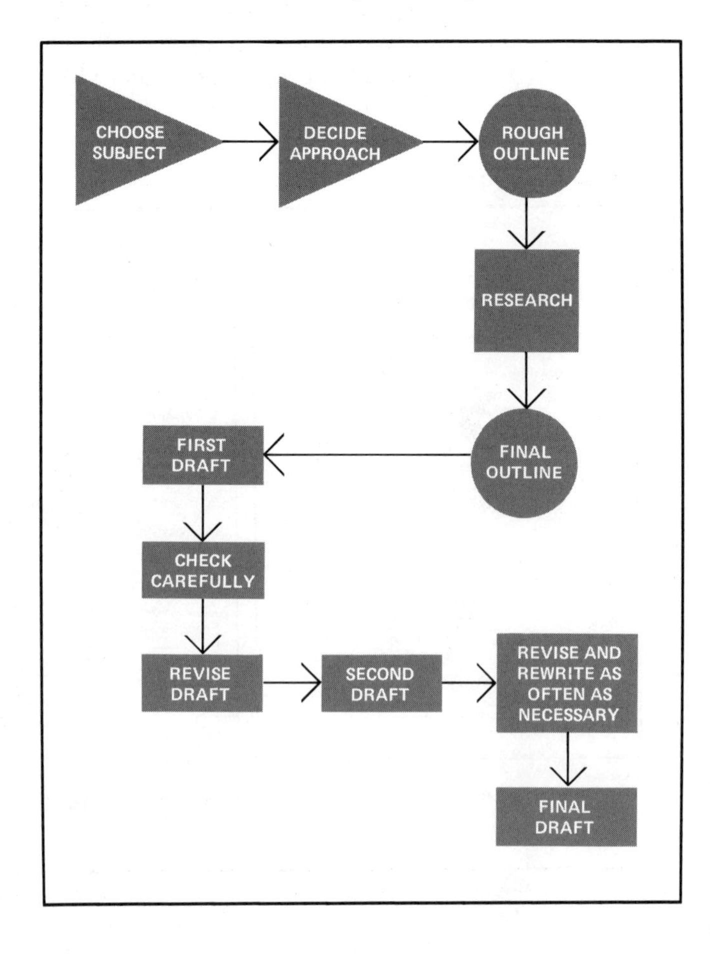

FIGURE 6-11. FLOW CHART OF THE PROCESS OF WRITING

FIGURE 6-12. SCHEMATIC DIAGRAM OF V-F CONVERTER

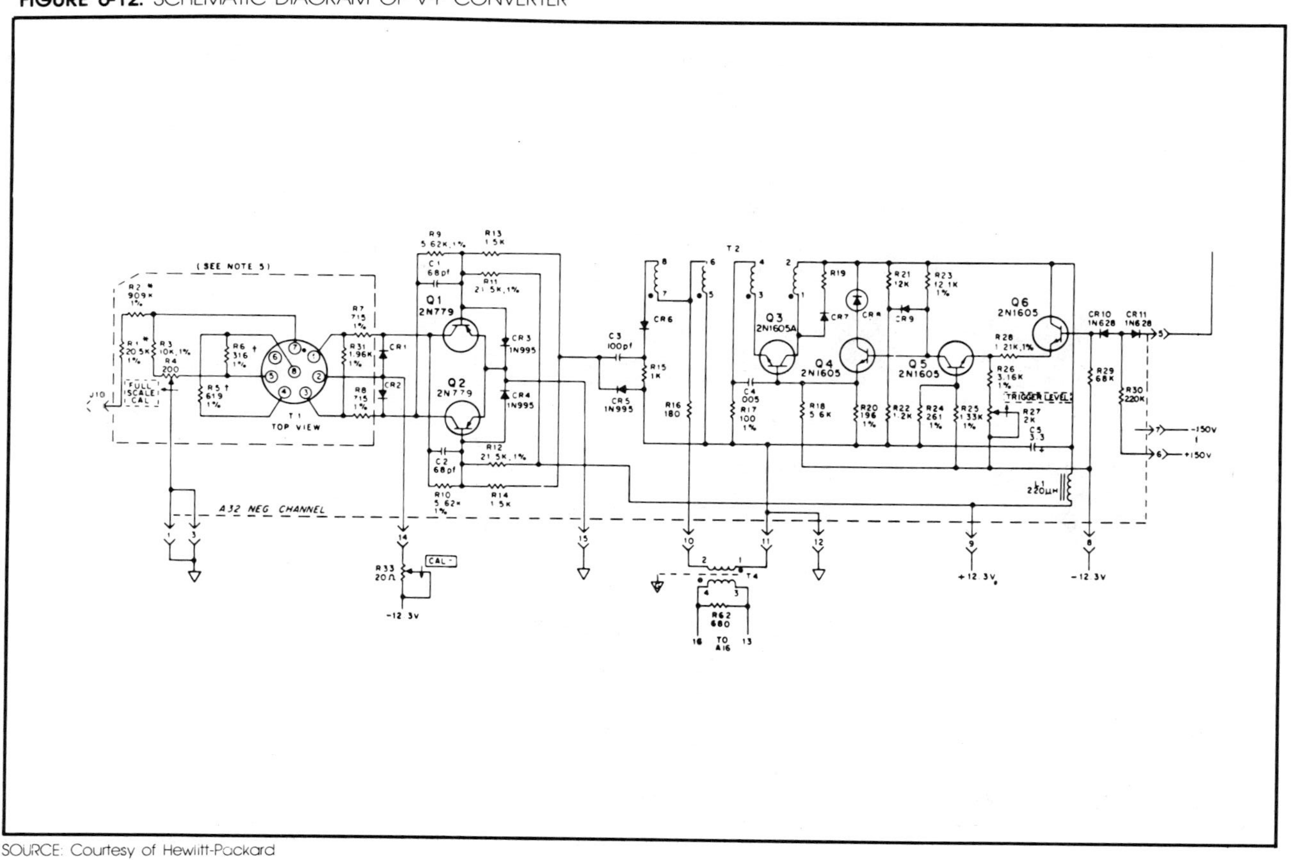

SOURCE: Courtesy of Hewlitt-Packard

FIGURE 6-13. ORGANIZATIONAL CHART OF A TYPICAL COLLEGE

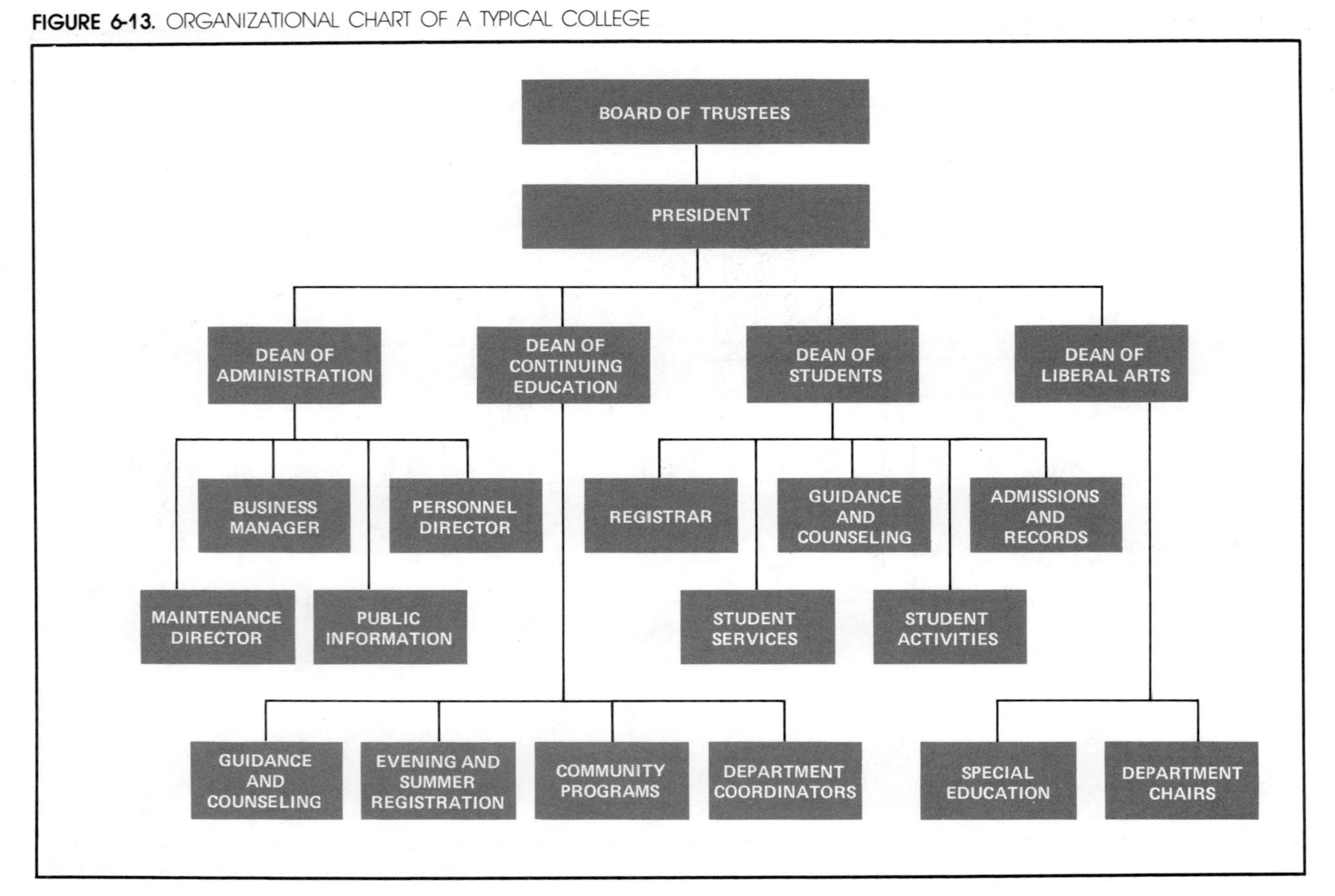

FIGURE 6-14. EXPLODED-VIEW DRAWING OF THE STOP GEAR REPLACEMENT FOR HP MODEL G, H, AND J532A

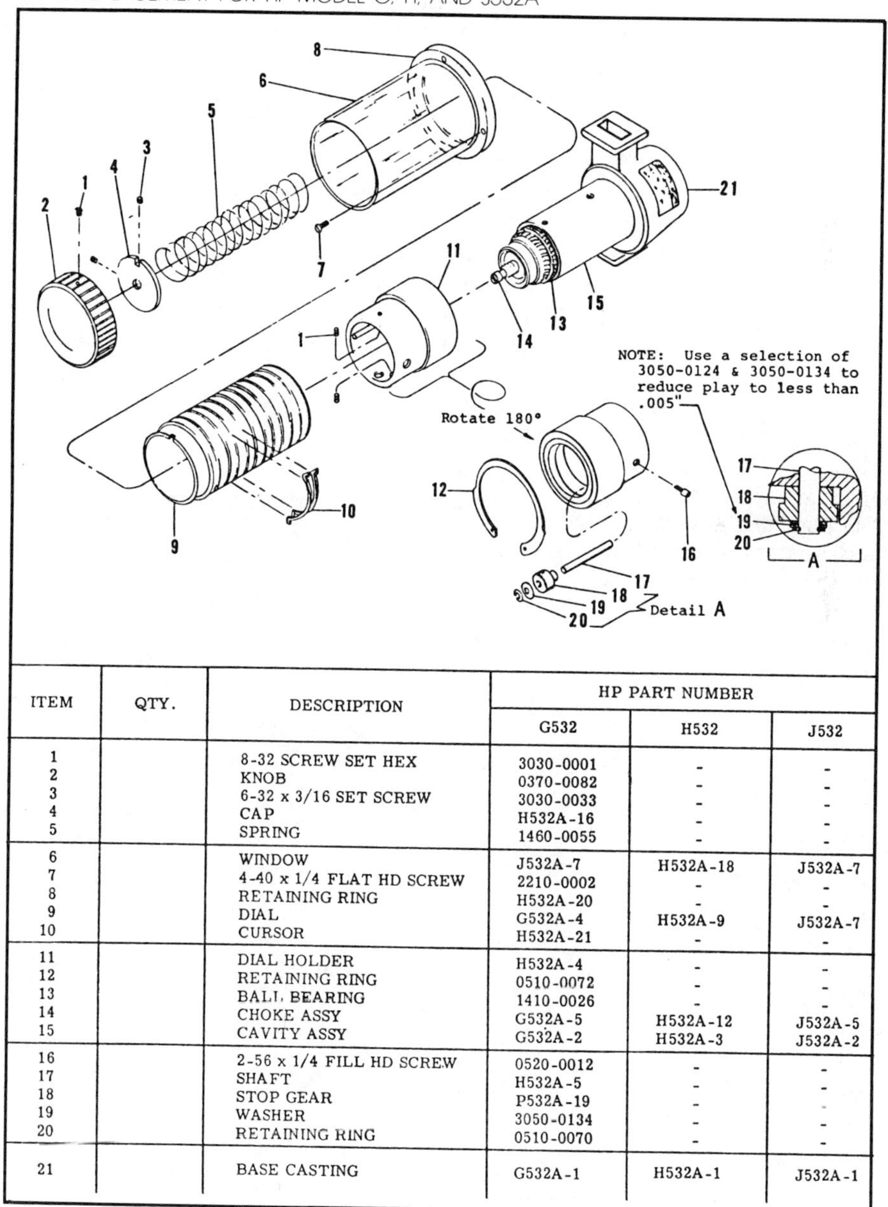

ITEM	QTY.	DESCRIPTION	HP PART NUMBER		
			G532	H532	J532
1		8-32 SCREW SET HEX	3030-0001	-	-
2		KNOB	0370-0082	-	-
3		6-32 x 3/16 SET SCREW	3030-0033	-	-
4		CAP	H532A-16	-	-
5		SPRING	1460-0055	-	-
6		WINDOW	J532A-7	H532A-18	J532A-7
7		4-40 x 1/4 FLAT HD SCREW	2210-0002	-	-
8		RETAINING RING	H532A-20	-	-
9		DIAL	G532A-4	H532A-9	J532A-7
10		CURSOR	H532A-21	-	-
11		DIAL HOLDER	H532A-4	-	-
12		RETAINING RING	0510-0072	-	-
13		BALL BEARING	1410-0026	-	-
14		CHOKE ASSY	G532A-5	H532A-12	J532A-5
15		CAVITY ASSY	G532A-2	H532A-3	J532A-2
16		2-56 x 1/4 FILL HD SCREW	0520-0012	-	-
17		SHAFT	H532A-5	-	-
18		STOP GEAR	P532A-19	-	-
19		WASHER	3050-0134	-	-
20		RETAINING RING	0510-0070	-	-
21		BASE CASTING	G532A-1	H532A-1	J532A-1

SOURCE: Courtesy of Hewlitt-Packard

Drawings and Diagrams. Drawings and diagrams help to clarify the description of complicated objects or processes. Both take various forms. A conventional drawing represents an object; a *cutaway* drawing shows the internal parts of an object as though the object were sliced in half; an *exploded-view* drawing (see Figure 6-14) shows the parts of an object individually and in sequence. An exploded-view drawing need not be drawn to scale, since its function is to show all the parts of a unit rather than their precise measurements (which, if desired, can be included in a listing of the parts).

A *blueprint* is a diagram that shows the detailed, working plan of a structure; an *assembly plan* is a

FIGURE 6-15. MAP OF THE UNITED STATES, SHOWING CENSUS DIVISIONS AND REGIONS

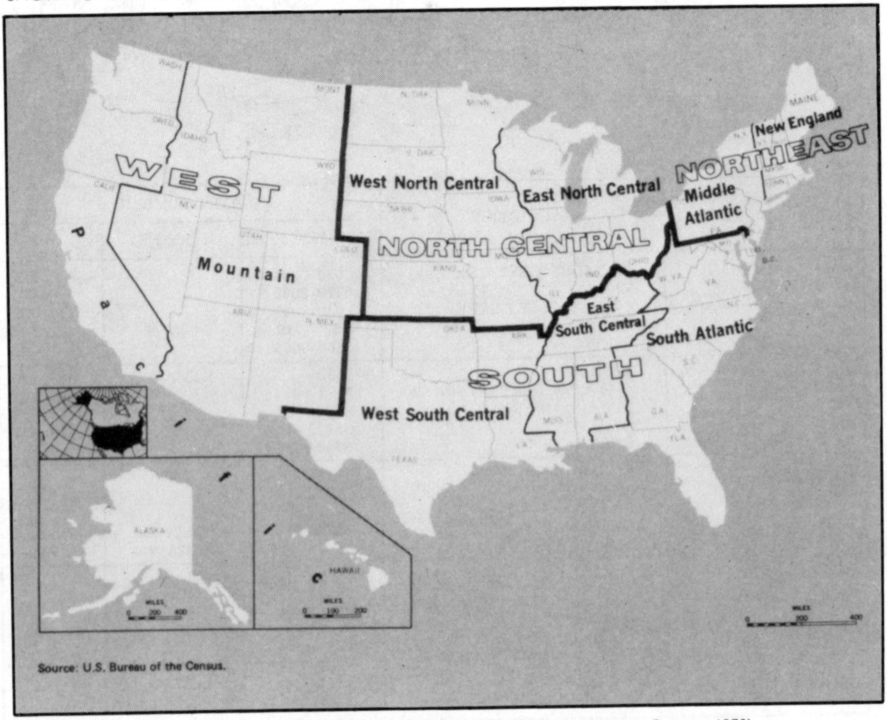

Source: U.S. Bureau of the Census.

SOURCE: Statistical Abstract of the United States (Washington, D.C. U.S. Bureau of the Census, 1978).

diagram for assembling the parts of a unit. Special maps, such as the one in Figure 6-15, are used to show patterns of distribution and concentration.

All relevant parts of a drawing or diagram should be clearly labeled. The choice of illustration, the amount of detail shown, and the clear identification of parts are the writer's responsibility, no matter who does the artwork.

6-7 SPELL OUT APPROXIMATE NUMBERS, NUMBERS UNDER 10, AND NUMBERS THAT BEGIN A SENTENCE; OTHERWISE USE NUMERALS.

Paradoxically, the attempt to achieve consistency in the use of numerals in technical writing has produced inconsistencies of its own. For example, some stylists prefer 10 as the dividing line between written numbers and numerals, others 20; some prefer a hyphen with mixed numbers (5-1/6), others omit it (5 1/6). These discrepancies, however, are not important. What is important is to be consistent within each piece of writing.

Here the use of 10 as the basic dividing line is suggested because numerals stand out more than written numbers: they provide instant recognition for the reader. And since numbers may appear often in technical writing, as many as possible should be instantly recognizable.

The use of a hyphen with mixed numbers also is suggested here, for similar practical reasons. Most typewriters do not have keys for all fractions, so a mixed number, say, 5-1/6, must be typed with the *5, hyphen, 1, slash,* and *6* keys. Without the hyphen (5 1/6), the 5 and 1 could be mistaken in a quick reading for 51. So the hyphen helps protect against misunderstanding.

Aside from such discrepancies, the problem most often faced by writers is how to mix numbers under and over 10. On this the standard is consistent: the larger number, or a mixed number (2-3/8), determines the form, which is numerals. And the unit of consistency may be the sentence, paragraph, or page.

Here the paragraph is suggested as the unit of consistency on two grounds. One, the paragraph is the basic unit of organization. And two, in using the paragraph instead of the sentence or page, you steer a middle course between fluctuation and inflexibility.

Some additional conventions in the use of numerals are listed below:

1. Express units of measurement in numerals: *110-volts, 8 meters, 7 percent, 90 rpm, 6 gallons, 3kHz, $2.*

2. When one number is followed immediately by another, write one of them out—usually the first, but sometimes the smaller: *eight 2-inch brackets, 110 six-foot boards.*

3. Use numerals in lists and references: *Table 7, Figure 3, page 9, Chapter 5.*

4. Write out fractions, but use numerals for mixed numbers: *three-fourths, 7-1/4.*

5. Write out street numbers under 10, but use numerals for all highway numbers: *Ninth Avenue, U.S. Interstate 5.*

6. Place a zero before a decimal so it won't be mistaken for a whole number: *0.105.*

7. Omit the comma from a four-digit number: *4873.*

6-8 PLACE FOOTNOTES AT THE BOTTOM OF THE PAGE AND A BIBLIOGRAPHY AT THE END OF A REPORT OR PAPER.

Footnotes

Information that is necessary but would interrupt the flow of the text may be added as a footnote. Footnotes either comment on or further explain the text, or name the specific source of material, especially quoted material.

When using footnotes, place a number in the text at the end of the material it comments on, and repeat the number at the bottom of the page, as shown here.[1] Use as many numbers as there are separate notes, and start with 1 again with the next section or chapter.

Where numbers would be confusing—as in tables or

[1]Footnotes also can be accumulated and listed, chapter by chapter, at the end of a book, paper, or report.

with mathematical formulas—use an asterisk (*) instead of a number. Follow it with a dagger (†), then a double dagger (‡), then with two asterisks (**), two daggers (††), and so on.

Several examples of footnotes that comment on the text appear in this book. Examples of footnotes that cite sources appear on the following pages.[2] Most source material comes from books, periodicals, newspapers, or other reports. The purpose of a source citation is to enable a reader to find the original source, wherever it is, whether in the library, or in an agency or company file.

The name of the author is the most direct path to the source and the first of the four elements of a reference listing. The other three elements are: the title of the original book, periodical, newspaper, or report in which the author's work appears; the publishing history of the work; and the volume and page number of the cited material. Where books are concerned, the edition, if there is more than one, also must be listed. All these elements are considered in order below.

For the first listing of a book, use the following format:

1. Begin with the author's full name, as printed on the title page. Use the first name first, and follow the last name with a comma. Do not use initials unless the author uses them. Where there are two authors, list them both; where there are three, list the first and add "et al." (from the Latin, *et alii,* "and others").

If the material in the book has been compiled by an editor or editors, list the name(s) of the editor(s) and add a comma and the abbreviation "ed." or "eds." If the editor also has translated the book, use the abbreviations "ed. and trans." If a different person has translated

[2]Reference footnotes may be eliminated entirely by inserting into the main text, in place of the footnote number, parentheses containing the last name of the author of the cited text and the year of publication: (Freud, 1905). The full citation is then listed as part of the bibliography. This is common practice in research and technical papers.

the book, add that person's name, a comma, and the abbreviation "trans." Where the book is an edited or translated version of a single author's work, place the author's name first and the editor's or translator's name *after the title.* When neither author, editor, nor translator is given, use the title of the book; avoid using "Anonymous" or "Anon."

Place a comma after the author's name. Where two authors are listed, use a comma after the second name only. Also use a comma after the abbreviations "et al."; "ed." or "eds."; or "trans."—but omit this comma where "ed.," "eds.," or "trans." appears *after* the title. Several types of authors' listings appear in the complete citations below:

[1] Joseph A. Alvarez, The Elements of Technical Writing (New York: Harcourt Brace Jovanovich, 1980), p. 126.

[2] Francis Christensen and Bonniejean Christensen, Notes Toward a New Rhetoric: Nine Essays for Teachers, 2nd ed. (New York: Harper & Row, 1978), p. 79.

[3] Charles T. Brusaw et al., Handbook of Technical Writing (New York: St. Martin's Press, 1976), p. 157.

[4] Donald H. Cunningham and Herman A. Estrin, eds., The Teaching of Technical Writing (Urbana: National Council of Teachers of English, 1975), p. 117.

[5] Artistotle, The Rhetoric of Aristotle, trans. Lane Cooper (New York: Appleton-Century-Crofts, 1932), p. 8.

[6] A Manual of Style, 12th ed. (Chicago: University of Chicago Press, 1969), p. 351.

2. Add the title of the book, as printed on the title page, and underline it to indicate italics. If the title includes a subtitle, place a colon after the title and follow it with the subtitle (also underlined). Capitalize the initial letter of the first word of the subtitle. Place a comma after the title (or subtitle) *only* if it is followed by an edition or volume number ("2nd ed." or "2 vols.") or by the abbreviation "trans."; otherwise, use no punctuation before the parentheses that enclose the publishing history.

3. Note the edition, if it is any but the first, and the number of volumes, if there are more than one. Place either or both after the title (or subtitle, if there is one) and before the publishing history. Use Arabic numerals and abbreviations—"3rd. ed.," "2 vols.," "rev. ed.," and so on. When noting both edition and volumes, place the edition number first, and follow it with a comma (but use no punctuation after the number of volumes).

4. Add the publishing history of the book, which consists of the city of publication, the publisher, and the date of publication. Enclose the complete publishing history in parentheses and follow it (but do not precede it) with a comma. Within the parentheses, place a colon after the city of publication and a comma after the publisher. Where a city is unfamiliar, add the abbreviation of the state to the listing: Englewood Cliffs, N.J.

Take the city of publication and the name of the publisher from the title page of the book, and the date of publication from the copyright page. Even if the publisher's name has changed (for example, through merger with another publisher) since the book was published, use the name on the title page. Where no city of publication, or no publisher, is given, use "n.p."; where no date is given, use "n.d."

5. End the footnote with the volume number (if there is one), the page number that contains the material, and a period. If *both* a volume and page number are required, omit the abbreviations "vol." and "p." Use instead Arabic numerals separated by a colon and followed by a period. For example, vol. 2, p. 213. becomes 2:213. But where *only* a page number is involved, retain the abbreviation: p. 246. Where more than one page is involved, use the abbreviation "pp.": pp. 15-17.

When listing standard reference works, omit the publishing history (but include the edition if it is not the first). Refer to the specific entry with the abbreviation "s.v.," for the Latin *sub verbo,* "under the word":

[1] The New Columbia Encyclopedia, 3rd ed., s.v. "trichina."

For repeat listings of a book in footnotes, there are two formats—the first short, the second shorter:

1. List the last name of the author, a shortened title of the book (underlined to indicate italics), and the page number. Separate each element by a comma, and follow the whole with a period. For example:

[4] Alvarez, Technical Writing, p. 162.

2. List only the last name of the author, followed by a comma, the page number, and a period:

[4] Alvarez, p. 182.

Either format is acceptable, but do not mix them. Also avoid the use of "op. cit." and "loc. cit." for repeat listings of books. Abbreviations of the Latin *opera citato,* "in the work cited," and *loco citato,* "in the place cited," these notations are more confusing than helpful.

For consecutive listings of the same book, use "Ibid." (an abbreviation of the Latin *ibidem,* "the same place"), followed by a comma, the page number, and a period. If even the page number is the same, simply use "Ibid.":

[3] Mario Pei, The Story of English (New York: Fawcett Publications, 1962), p. 87.

[4] Ibid.

[5] Ibid., p. 103.

Footnote 5 indicates that the reference is identical except for the page number.

The format for periodicals is similar. It varies only in the listing of the title and the publishing history—here the volume and issue numbers and the date of publication. Place a comma and the title of the article after the author's name, and after it another comma and the title of the journal. Enclose the article title in quotes (with the final comma inside the second quote mark), and underline the journal title to indicate italics. Treat the publishing history as follows:

1. Place the volume number directly after the journal title (without an intervening comma). If each issue of a volume is paginated consecutively throughout, omit the issue number. Follow instead with the year of publication (in parentheses), a colon, the page number, and a period:

> [1] Mary Edel Denman, "The Measure of Success in Writing," College Composition and Communication 29 (1978): 43.

Note that, as in the listing of books, the volume number is written in Arabic rather than Roman numerals for clarity, and that the abbreviations "vol." and "p." are omitted.

2. If each issue of a volume is not paginated consecutively, cite the issue number after the volume number, preceded by a comma and the abbreviation "No." Follow the issue number with the year of publication (in parentheses), a colon, the page number, and a period:

> [1] H. Lee Shimberg, "Ethics and Rhetoric in Technical Writing," Technical Communication 25, No. 4 (1978):16.

3. If only the issue number and a date but no volume number are given, cite the issue number after the journal title, preceded by a comma. Follow the issue number with the date (in parentheses), a *comma* (instead of a colon), the abbreviation "p.," the page number, and a final period:

> [1] H. Lee Shimberg, "Ethics and Rhetoric in Technical Writing," Technical Communication, No. 4 (1978), p. 16.

The "p." above is retained because it is only dropped when the "vol." is dropped, and here there is no "vol." to drop.

4. When only the date of a magazine is given, or only the date is essential, place it after the magazine title:

[1] Randolf Menzel and Jochen Erber, "Learning and Memory in Bees," Scientific American, July 1978.

Note in the date the absence of a comma between the month and year.

For *consecutive* listings of periodicals in footnotes, use the abbreviation for the consecutive listings of books: Ibid., and add the page number if it is different. For repeated but *not* consecutive periodical listings, use only the author's last name, the title of the *article* (shortened when possible), the page number, and a final period:

[4] Denman, "Measure of Success," p. 45.

[5] Menzel and Erber, "Learning and Memory," p. 102.

[6] Ibid., p. 104.

[7] Ibid.

[8] Denman, "Measure of Success," p. 46.

When citing a newspaper, use the name of the paper (underlined to indicate italics), the date of publication, and the page number. Separate each element with a comma, and end the citation with a period:

[1] San Francisco Chronicle, 18 November 1977, p. 3.

Note that in the date the day is placed before the month. The name of the month may be abbreviated if it is done throughout the listings.

When the source is another report, or unpublished

material (such as a dissertation), use the following format:

1. Begin with the author's full name, first name first, followed by a comma after the last name.
2. Add the title of the report and enclose it in quotes if the report (or dissertation) has not been published. Underline to indicate italics *only* if the material has been published.
3. Name the source of the report. Where it is a company, place a comma and the name of the company after the title, followed by another comma and the city in which the company is headquartered. Add the date of the report (in parentheses), a comma, the file number (if available), the page number, and a period. Where the source is a university, as with a dissertation, use the abbreviation "Ph.D. diss.," followed by a comma, the name of the university, another comma, and the year of publication—all enclosed in parentheses, and followed by a comma, the page number, and a final period. An example of each format appears below:

> [1] Carole Bendel, "Stress Reduction Studies Among Teledex Employees," Teledex, Inc., Palo Alto (October 1977), No. 83207, p. 28.
>
> [2] Edward Bannaman, "The Role of Perception in Phenomenological Psychology" (Ph.D. diss., Northwestern University, 1975), p. 218.

A final word about footnotes: their precise format appears intimidating. But it will seem less so if you think in terms of the four parts of a footnote—author, title, publishing history, and volume and page number—and cite each part in turn.

Bibliographies

Unlike footnotes, which may explain or comment on the text or indicate reference sources, a bibliography only lists reference sources. **(See also 8-2, p. 158.)** These usually are those sources cited in individual footnotes (or in parenthetical references inserted into the text). All the sources cited in parenthetical references must be covered in a bibliography. This does not apply, however, to sources cited in reference footnotes. A bibliography may cover all footnote sources, or only the significant or frequently cited ones; the choice lies with the writer.

Unlike footnotes, bibliographies are arranged alphabetically according to the author's last name (or, lacking an author's name, according to the title of the work). The format of a bibliography is similar to that of the reference footnotes, with a few variations. These are mainly the transposing of the author's name, the separation of the author's name from the title by a period instead of a comma, and the separation of the title from the publishing history by a period instead of parentheses (which are dropped altogether).

Each variation is described in detail below:

1. In a bibliography transpose the author's name from first name first to last name first; add a comma after the last name and a period after the first. Where there are two authors, transpose the names of both and separate them with the word "and," followed by a comma. Where there are three or more authors, instead of using the abbreviation "et al.," list every name, transposed, and separate each with a semicolon, adding the word "and" after the final semicolon. Always separate the final name from the title by a period.

Where more than one work is credited to an author (or authors), substitute a line for the author's name in all entries after the first, and list the works alphabetically by title.

If the work is a book in which the material has been compiled by an editor or editors, follow the name(s) of the editor(s) with a comma and the abbreviation "ed." or "eds." Where an editor or translator's name appears in addition to an author's name, avoid abbreviations. Use "Edited by" or "Translated by," followed by the appropriate name(s) and a period. Such citations should follow the title of the work and precede the publishing history of it, separated from both by periods.

Finally, if neither author, editor, nor translator is given, list the book according to title.

Several types of authors' listings appear in the complete listings below:

A Manual of Style. 12th ed. Chicago: University of Chicago Press, 1969.

Alvarez, Joseph A. The Elements of Technical Writing. New York: Harcourt Brace Jovanovich, 1980.

Aristotle. The Rhetoric of Aristotle. Translated by Lane Cooper. New York: Appleton-Century-Crofts, 1932.

Brusaw, Charles T.; Alred, Gerald J.; and Oliu, Walter E. Handbook of Technical Writing. New York: St. Martin's Press, 1976.

Bryant, Margaret M. Current American Usage. New York: Macmillan, 1962.

———. Modern English and Its Heritage, 2nd ed. New York: Macmillan, 1962.

Christensen, Francis, and Christensen, Bonniejean. Notes Toward a New Rhetoric: Nine Essays for Teachers, 2nd ed. New York: Harper & Row, 1978.

Cunningham, Donald H., and Estrin, Herman A., eds. The Teaching of Technical Writing. Urbana, Ill.: National Council of Teachers of English, 1975.

Note that in bibliographical listings all lines after the first are indented so that the name of the author stands out. Also note the differences between some of the citations above and the same citations used to illustrate reference footnotes on p. 126.

2. As shown in the examples above, in a bibliography separate the title from the publishing history by a period, and drop the parentheses that enclose the publishing history of a book in a footnote. Where periodical titles are involved, separate the title of the article from the title of the journal by a period (placed inside the quote marks) instead of a comma:

Denman, Mary Edel. "The Meaning of Success in Writing." College Composition and Communication 29 (1978): 42-46.

Note that in a bibliography the total number of pages in an article is listed rather than a specific page number corresponding to a specific reference in a footnote.

Scientific bibliographies vary even further. The format of a scientific bibliography is distinct from that of a regular bibliography in three respects. These are: the listing of the author's name, of the title of the work, and of its publishing history. Each variation is described below:

1. In a scientific bibliography place the author's last name first, followed by a comma—but use initials only for first and middle names: Wilson, E. O.

2. In a scientific bibliography capitalize the words of the title as you would a regular sentence:

The theory of island biogeography.

Lemur behavior: a Madagascar field study.

Where the reference is to a journal article, the article title may be omitted entirely when it is done throughout the bibliography. If the article title is used, however, omit the usual quote marks around it, and abbreviate the *journal* title:

The dynamics of learning in the honeybee. J. Comp. Phys.

J. Comp. Phys. is the abbreviation for *Journal of Comparative Physiology.* (For standard abbreviations of journal titles, consult *Chemical Abstracts List of Periodicals* [1961].)

3. In a scientific bibliography place the year of publication after the author's name and before the title of the work, separated from each by a period. Place the rest of the publishing history at the end of the listing—but abbreviate the publisher's name:

Selye, H. 1956. The stress of life. New York: McGraw.

Partridge, R. B. 1969. The primeval fireball today. Amer. Sci. 57: 37-74.

Partridge, R. B. 1969. Amer. Sci. 57: 37-74 [article title omitted].

Amer. Sci. is the abbreviation for *The American Scientist;* "McGraw" is the abbreviation for "McGraw-Hill Book Company." (For standard abbreviations of book publishers' names, consult *Books in Print.*)

Finally, in a bibliography, as in a footnote, take the author's name, the title, the publisher's name, and the date and place of publication from the reference source itself.

For further information on footnotes and bibliography, consult the latest edition of *A Manual of Style,* published by the University of Chicago Press.

For illustrative purposes, a selected bibliography (a list of the main sources) for this book appears below rather than in an appendix.

A Manual of Style. 12th ed. Chicago: University of Chicago Press, 1969.

Allen, Robert L. English Grammars and English Grammar. New York: Charles Scribner's Sons, 1972.

Brusaw, Charles T.; Alred, Gerald J.; and Oliu, Walter E. Handbook of Technical Writing. New York: St. Martins Press, 1976.

Bryant, Margaret M. Current American Usage. New York: Funk & Wagnalls, 1962.

———. Modern English and Its Heritage. 2nd ed. New York: Macmillan, 1962.

Christensen, Francis, and Christensen, Bonnijean. Notes Toward a New Rhetoric: Nine Essays for Teachers. 2nd ed. New York: Harper & Row, 1978.

Cunningham, Donald H., and Estrin, Herman A., eds. The Teaching of Technical Writing. Urbana: National Council of Teachers of English, 1975.

Gates, Jean K. Guide to the Use of Books and Libraries. 3rd ed. New York: McGraw-Hill, 1974.

Hall, Donald. Writing Well. 2nd ed. Boston: Little, Brown, 1976.

Houp, Kenneth W., and Pearsall, Thomas E. Reporting Technical Information. 3rd ed. Beverly Hills, Calif.: Glencoe Press, 1977.

Jesperson, Otto. Growth and Structure of the English Language. 9th ed. New York: Macmillan, 1938.

Miles, Robert, and Bertonasco, Marc. Prose Style for the Modern Writer. Englewood Cliffs, N.J.: Prentice-Hall, 1977.

Mills, Gordon H., and Walter, John A. Technical Writing. 4th ed. New York: Holt, Rinehart and Winston, 1978.

Sparrow, W. Keats, and Cunningham, Donald H. The Practical Craft: Readings for Business and Technical Writers. Boston: Houghton Mifflin, 1978.

Strunk, Jr., William, and White, E. B. The Elements of Style. 3rd ed. New York: Macmillan, 1979.

Ulman, Jr., Joseph N., and Gould, Jay R. Technical Reporting. 3rd ed. New York: Holt, Rinehart and Winston, 1972.

Willis, Hulon. A Brief Handbook of English. New York: Harcourt Brace Jovanovich, 1975.

EXERCISES

1. Write an abstract of Section 1 of this book.

2. Write a summary of Section 1 of this book.

3. Write a resumé and covering letter for a job that you feel you would qualify for.

4. Using any product that is advertised, test the product against the advertising claims. Then write a letter report to the manufacturer of the product, describing your test and your conclusions regarding the validity of the advertising claims. Make any advisable recommendations.

5. Write a one-page memo or letter to the appropriate person suggesting something specific you would like to see done (or abolished) in your town, school (or class), or job. The choice of letter or memo, and the tone you use, should reflect your relationship to the individual to whom the message is directed.

6. Check the following paragraph for consistency in the use of numbers. If necessary, change any numerals to spelled-out numbers, or vice versa:

> A study of the New York Times shows that in the last one hundred years punctuation marks per sentence have decreased from 3.39 to 2.55. The use of the comma alone has decreased by almost thirty-three percent. Lincoln's Gettysburg Address, for example, contains thirty-eight punctuation marks in its ten sentences, or 3.8 per sentence. Twenty-one of these are commas, but about half of them would not be used today. Dropping ten of them would bring the average punctuation down to 2.8 per sentence, closer to today's figure.

7. Obtain from your local weather bureau the temperature ranges for your area over any 30-day period. Using this data, compile a table for the high, low, and average temperatures and show the trend of each in one line graph.

8. Compile a bibliography of the books you use in your various classes. Then list the same books as footnotes. Note the differences in format.

Strictly speaking, style is the selection and arrangement of words. But that is only its technical dimension. The basic dimension, the one that governs the selection and arrangement, is who and what we are. In short, style is the person. Everything that has shaped us—our genes, the people in our lives, the world around us, and our experiences—shapes every sentence we write. If few of us see this, it is because few of us have been trained to.

A classic example of style as the personification of the writer is Julius Caesar's description of his swift victory over the Gauls, the ancient people of France. *Veni, vidi, vici*, he wrote, which translates: "I came; I saw; I conquered." One could as easily write: "I arrived at the battlefield; I beheld the enemy; I won the victory." The message is the same; but missing is Caesar's air of supreme self-confidence, even bravado. Also missing is the sense of the swiftness of the victory.

The poet Donald Hall notes that a change in style, however slight, "is always a change in meaning, however slight." We see that in the comparison above. Caesar's version implies that he personally was responsible for the swift victory. The paraphrase misses both the bravado and the swiftness—and therefore part of the meaning of Caesar's message. Style, then, not only reflects the writer but shapes the message accordingly.

7-1 BE SELECTIVE: FOCUS ON THE ESSENTIAL INFORMATION, THE SIGNIFICANT DETAIL.

Selectivity, the first step to an effective style, depends upon knowledge. You must know: (1) a report's purpose and audience; (2) enough about the subject of the report; and, most important, (3) what you want to say in the report.

Ironically, the more you know about a subject, the harder it is to write about it selectively. That's why it's so important to know what you want to say: it provides a standard for selecting. Knowing what you *want* to say helps you decide what you *must* say—which leads you to the essential information, the significant detail.

When a choice is not clear-cut, choosing becomes more difficult because we know we will lose almost as much as we gain—perhaps even lose more than we gain, if we choose poorly. Loss is the price of judgment. Still, we must choose, must be selective, for to write exhaustively is to exhaust the reader. To say *everything* is to lose the essential in a snowfall of words.

7-2 DEVELOP A LEAN, DIRECT STYLE; AVOID INFLATED LANGUAGE AND RAMBLNG SENTENCES.

A sentence is shaped as is a sculpture: by building it up or paring it down. Each technique produces sentences distinct in form and content—and effect. The built-up sentence itself comes in at least three varieties. One consists of separate, related concepts combined in one or more expanded clauses, the longer the better; another contains an overload of adjectives, adverbs, and verb phrases. Avoid these; one is lifeless, the other inflated.

Francis Christensen, who for many years taught writing at the University of Southern California, suggested a third, which he called the *cumulative* sentence. It consists of a main clause (which may be preceded by a modifying word or phrase), followed by additional word clusters such as subordinate clauses and noun and verbal phrases.

In the following extract, Christenscn uses a cumulative sentence to explain the effect such sentences create. Each of the additions has been numbered for reference below:

> The main clause advances the discussion; but the additions move backward, (1) as in this clause, (2) to modify the statement of the main clause or more often explicate or exemplify it, (3) so that the sentence has a flowing and ebbing movement, (4) advancing to a new position and then pausing to consolidate it, (5) leaping and lingering as the popular ballad does.

The main clause may be preceded by a modifying word or phrase, but in Christensen's cumulative sentence above it is not. His sentence instead contains two main clauses: *The main clause advances the discussion,* and *the additions move backward.* (The conjunction *but* connects the two clauses.) The entire sentence consists of two clauses: the first main clause, and the second main clause plus all the additions.

By *additions* Christensen means everything that follows the main clause. In his sentence it is the second main clause, which is followed by five additions, each separated from the others by commas. When he says "the additions move backward," he means that each addition modifies, explains, or exemplifies the statement in the main clause (or in a prior addition). This produces, simultaneously, a double effect. It subtly directs the reader's attention back to the main clause (or prior addition) even as that attention flows forward in the direction of the expanding sentence. The result is "a flowing and ebbing movement" in the sentence.

For example, the first addition *(as in this clause)* explains that the second clause itself is an example of what Christensen is describing in the sentence. The second addition *(to modify...it)* explains why the addition moves backward. The third addition *(so that...movement)* explains the effect of the modification produced by the prior addition. And the fourth and fifth additions are explanations of the "flowing and ebbing movement" mentioned in the third addition.

The cumulative sentence, as Christensen demonstrates, adds breadth and subtlety to expression. It enables the writer to focus on a subject in a main clause and then, like a camera, pan around that subject, exploring its various aspects to produce a complex, in-depth picture of it. This technique is often used in fiction and essays to explore the subtleties of character or concept. But it also is effective in technical writing—for example, in describing the implications of theory or policy, the multiple facets of personality, or the subtleties involved in decision making.

In the description of objects and processes, however, where the basic thrust is forward rather than back and forth, use the pared-down, lean sentence. Its simplicity, directness, and conciseness provide the clarity and drive that technical description often lacks. Figure 7-1 below illustrates how a lean sentence is sculpted.

FIGURE 7-1.
PARING DOWN AN INFLATED SENTENCE

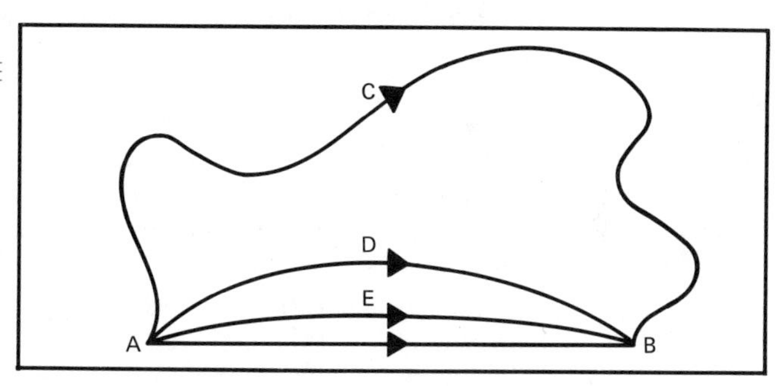

Line AB represents the ideal sentence: the straight line that marks the shortest distance between two points. Line ACB represents a rambling, inflated sentence—the kind of sentence too often found in technical writing. The message is there somewhere, but it is obscured by the words used to convey it:

> It is my definite impression that this new system would be of benefit to the critical care units in our delivery of care and in our efforts to give the best possible treatment to our patients.

Thirty-six words, but little clarity or drive.

Line ADB represents the same sentence stripped of

some of its inflated language and digressions:

It is my impression that this new system would help us give the best possible treatment to our patients in the critical care units.

Twenty-four words, and now clearer and more direct. An important consideration—that the system would affect the critical care units—also has been shifted from the middle of the sentence, where it was obscured, to the end, a position of emphasis.

Perhaps the sentence can be made even more direct. Line AEB represents an attempt to pare it down further:

I believe this new system would help us provide optimum treatment of patients in the critical care units.

Eighteen words now, half as long as the original: more concise as well as more direct. The main difference is the substitution of *I believe* for *It is my impression that.* Replacing the bland verb *give* with the precise verb *provide* also adds accuracy to the description. And the single word *optimum* does the work of the phrase *the best possible.*

Other changes might produce an even more direct sentence. Human fallibility, however, prevents us from achieving the ideal sentence AB; the art lies in coming as close to it as we can.

7-3 WRITE IN THE PRESENT TENSE WHENEVER POSSIBLE FOR SIMPLICITY AND IMMEDIACY.

Tense is the time machine of language. With it we move from the present backward to the past or forward to the future. And as we move, we cover distance; for as Albert Einstein showed in his theory of relativity, time is the fourth dimension, coordinate with length, width, and depth. Only in the present tense, then, is the action immediate to the reader. All other tenses, since they take us back to the past or forward to the future, move the action—however slightly—away from the reader, leaving between the reader and the action a gap—however small—that uncertainty or misunderstanding can fill.

The present tense is not only immediate but also one of the simplest and most direct of the six English tenses. These six are the present, past, and future (called the *imperfect* tenses), and the present perfect, past perfect, and future perfect (called the *perfect* tenses). Each tense also has a *progressive* form, a late development in English, which enables us to express an action while performing it: *I am writing*. The present tense *(I write)* does not express an ongoing action, only that writing is something I sometimes do. The difference may be seen in the sentence: *I am writing, but I don't always write.*

The six tenses and the progressive form of each are shown in Table 7-1.

TABLE 7-1. THE SIX ENGLISH TENSES AND THE PROGRESSIVE FORM OF EACH

IMPERFECT TENSES		
Tense	Regular Form	Progressive Form
PRESENT	I learn	I am learning
PAST	I learned	I was learning
FUTURE	I will learn	I will be learning
PERFECT TENSES		
PRESENT PERFECT	I have learned	I have been learning
PAST PERFECT	I had learned	I had been learning
FUTURE PERFECT	I will have learned	I will have been learning

Note that each perfect tense requires a form of the verb *have* and the past participle of the main verb *(learned)*. Also note that every progressive form requires a form of the verb *be* and the present participle of the main verb *(learning)*.

The perfect tenses, the last to develop in English, differ from the imperfect tenses in two respects. One, the perfect enables us to express a past action that continues into the present *(I have learned)* or goes further back into the past *(I had learned)* or is projected into the future *(I will have learned).* We cannot do this with the imperfect tenses. Two, this past and continuous action is always completed at a specific point in time, either in the present, the past, or the future.

Together the six tenses cover the spectrum of time from past through present to future. They can—and often do—cause confusion, however, because two pairs of them overlap in the part of the spectrum they cover, as Figure 7-2 shows. The pairs are the past and the present perfect, and the future and the future perfect. Although the past perfect does not overlap any other tense, it presents problems of its own. Only the present is without problems—which is one good reason to prefer it.

The problems of overlap and of the past perfect are discussed below and illustrated in Figure 7-2.

1. The past tense and the present perfect tense both describe actions in the past, but the past perfect describes past action that bears on the present. It bears on the present because its action is completed in the present—an indication also that the action itself occurred in the recent past. The statement "I learned to write by writing" tells us that the learning was in the past, but not exactly when in the past it was completed. In the statement "I have learned to write by writing," the learning also is in the past, but the action is completed in the present—the moment that the statement is made.

2. In the same way, the future tense and the future perfect tense both describe actions in the future. But the action described by the future perfect is completed at a specific point in the future, while the action described by the future is not. The statement "I will learn to write by writing" tells us only that the learning will occur sometime in the future; it doesn't say exactly when. But the statement "I will have learned to write by writing" is completed at a specific point in the future, even though no date is given. The specific point is that

moment when *I will have learned.* At that moment the learning will be over: it will be part of the past *(have learned).*

3. The past perfect does not overlap any other tense. It is used to describe a past event that occurred *before* another past event. The statement "I had learned to write by writing" could be expanded by "before I took your class," or by any other such statement requiring a verb in the past tense (no other tense would make sense). The past perfect here enables us to start from a point in

FIGURE 7-2. THE TIME SPECTRUM COVERED BY THE SIX ENGLISH TENSES

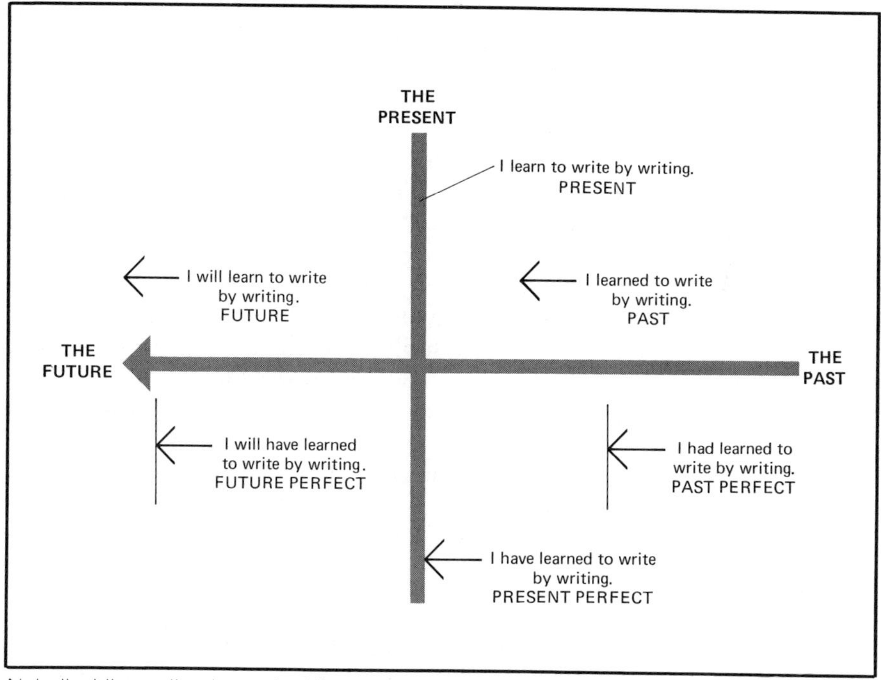

Note that the action in each of the perfect tenses is completed at a specific point in time, shown by the arrows touching the three vertical lines. In the present perfect it is the present: the action is completed the moment the writer makes the statement. In the past perfect it is a specific point sometime in the past. And in the future perfect it is a specific point somewhere in the future. Note also that the arrows for the past and future tenses do not touch any lines. This is meant to show that the action has not been completed at a specific point in time. Where the past is concerned, we know only that it was completed sometime before the present, but we don't know exactly when. Where the future is concerned, we know only that it will be complete in the future, but we don't know when.

the past (taking the class) and go further back into the past (the learning). And that starting point in the past is the specific point where the action of the past perfect is completed: the learning was completed *before I took your class.*

Related to the six verb tenses are the three *moods* of a verb: (1) the *indicative,* which makes a statement *(I learn);* (2) the *imperative,* which gives a command *(Learn!);* and (3) the *subjunctive,* which states a possibility, probability, doubt, wish, or contradiction of fact *(I may learn).* The subjunctive is sometimes called the *conditional*—a word that more clearly defines its function. It requires, along with the main verb, an *auxiliary,* or helping, verb: *may, might, could, would,* or *should.*

When any of the alternatives to the present tense are necessary to do the job, use it of course. But once in the present tense, do not change tenses except to show a real change in time. To change tenses for any other reason is to needlessly inject complexity and distance into what you write. Also avoid using the progressive form *(I am learning)* or the subjunctive mood *(I may learn)* unless they are necessary. Both require verb phrases, which are wordy and relatively weaker than solitary verbs. And the subjunctive adds to that drawback the weakness of being only conditional or probable.

7-4 USE EXAMPLES AND COMPARISONS TO CLARIFY DESCRIPTIONS AND EXPLANATIONS.

An example illustrates, models, or imitates whatever is being described or explained. A comparison points out similarities or differences between things. Comparisons that point out similarities between things otherwise unlike are called *similes, metaphors,* or *analogies.* Each differs slightly from the others.

A simile always contains the word *like* or *as:* "as unreliable as the weather." A metaphor is a compressed simile. It needs no signal such as *like* or *as* to announce the comparison; its strength lies in the image used: "No man is an island." An analogy is like a simile, but it may extend beyond a phrase or sentence: "In writing, knowing what you want to say is like walking on firm ground instead of on quicksand, or like exploring with a compass instead of without one."

Comparisons must be logically consistent—that is, compare things that are related, and measure them by the same standard. For example, compare a toaster to a blender but not to a telephone, a canal to a highway but not to a dam. Use the same units and standards in a comparison—not meters for one and decimeters for the other, nor the looks and personality of one person and the education and experience of the other.

Examples should be relevant and unambiguous—that is, illustrate only the point in question.

7-5 REPEAT WORDS OR PHRASES FOR CLARITY OR EMPHASIS, OR TO EASE TRANSITIONS; BUT AVOID NEEDLESS REPETITION.

Repetition, like most of what concerns us in writing, is neither right nor wrong but rather works or doesn't, depending on how it is used. How it is used—and when—should reflect the writer's judgment rather than carelessness.

Clarity

To avoid repeating a word needlessly either substitute a loose synonym or pronoun for it, or simply delete it. Note the sentence below:

> Most politicians agree that inflation is the nation's most serious problem, but few politicians are willing to act to control it.

The second *politicians* is needlessly repetitive. It could be replaced by a loose synonym, either *legislators, officials,* or *officeholders.* But *legislators* does not include all politicians—the president, for example. And *officials* includes appointees—bureaucrats, for example —who are not, strictly speaking, politicians. *Office-holders* is the tightest synonym for politicians, but it is bulky. More effective would be the substitution of the pronoun phrase *of them:*

> Most politicians agree that inflation is the nation's most serious problem, but few of them are willing to act to control it.

Also effective would be the deletion of the second *politicians:*

> Most politicians agree that inflation is the nation's most serious problem, but few are willing to act to control it.

In avoiding repetition, do not substitute a contrived variation, or an inaccurate one (such as *legislators* or *officials* would be in the above example). And do not replace or delete a repeated word (or phrase) when it would reduce clarity or destroy parallel structure. **(See also 3-7, p. 58.)** Note the sentences below:

> The tone of a letter may affect the reader more than its content—may even make or break the letter.
>
> The colon usually can be replaced by a comma or dash, the comma reducing emphasis and the dash increasing it.
>
> If few of us see this, it is because few of us have been trained to see it.

In the first sentence, replacing *the letter* with the pronoun *it* would reduce clarity: whether *it* referred to *letter* or *content* would be unclear. *The letter,* then, is repeated for clarity.

Deleting *the comma* and *the dash* in the second sentence would reduce clarity and break down the parallel structure of the noun phrase. *The comma* and *the dash,* then, are repeated for clarity and to maintain the parallel structure of the phrase. Note also that the final *it* substitutes for *emphasis,* thus avoiding unnecessary repetition.

In the third sentence, *see it* can be dropped without loss of clarity. This deletion also adds impact to the sentence because the final stressed syllable, which was *see,* becomes *trained,* the main idea of the sentence. Of course the sentence now ends with a preposition *(to)*;

and though some grammarians might insist on retaining *see it* to preserve form, the history of English shows that the users of the language prefer function over form when the two conflict. **(See also 3-7, p. 58.)**

Emphasis

Deliberate repetition, sometimes called *reiteration,* is one of the most effective devices for creating emphasis and rhythm in a sentence or paragraph. The rhythm, in fact, reinforces emphasis—which is why commercial jingles are so effective. Note the sentence below:

> Each form of the adjective had to agree with the form of the noun it modified—which itself had to agree with the form of the verb it combined with.

The use of *form* three times and *agree* twice emphasizes the basic idea of the sentence, which is that each word in a sentence (in Old English) had to agree in form.

Another effective device for creating emphasis and rhythm is *alliteration,* usually consecutive or alternate words beginning with the same letter or sound: *form follows function* or *earn and learn.*

Transition

A smooth transition is like the image of the Roman god Janus—it faces both forward and backward, while artfully linking one to the other.

A transition bridges the gap between two sentences or paragraphs; but whether sentences or paragraphs, the two should be related in some way. In this relationship the sentences or paragraphs may be similar, contrasting, coordinating, or repetitive; consecutive examples; or parts of a common group. Or one may be complementary or subordinate to the other, follow the other in logical sequence, or have a cause-and-effect or problem-solution relationship with the other. **(For examples of some of these relationships in sentences and paragraphs, see 4-3, p. 66.)**

TABLE 7-2. TRANSITIONAL WORDS AND PHRASES

SEQUENCE	PLACE	CONTRAST	SIMILARITY	CAUSE-AND-EFFECT	EXAMPLES	REPETITION
After	Above	Actually	Among	Accordingly	For example	Also
Afterward(s)	Across	Although	As	As a result	For instance	And
Before	Behind	Besides	Even	Because	Namely	As noted
During	Beside	But	In comparison	Consequently	That is	In fact
Earlier	Beyond	Conversely	Like	Furthermore		In short
Finally	Inside	Despite	Similarly	Hence		Indeed
First, Second, etc.	Near	Even though		If		Of course
Following	Outside	Fewer than		Since		
Later	Over	However		So		
Next	There	In contrast		Therefore		
Now	Toward(s)	Instead		Thus		
Once	Under	More than				
Then	Where	Nevertheless				
Until		Otherwise				
While		Rather				
		Unless				
		Yet				

Between such sentences or paragraphs, the relationship itself is the transition: the related idea or sentence structure or logical connection itself forms the link, bridges the gap. This is the most effective transition. But where such a relationship is lacking or is not strong enough, provide the link by repeating in the second sentence or paragraph a key word or phrase from the first one (as in the transition between this paragraph and the next).

A connecting word or phrase used at the beginning of the second sentence or paragraph may also serve as a transition. Most of these are conjunctions and prepositional phrases. They provide a transition when the connection involves either sequence, place, contrast, similarity, cause-and-effect, examples, or repetition itself (for clarity or emphasis). A list of such connectives, roughly classified, appears in Table 7-2. Avoid bulky phrases, however. Such constructions as "in addition to," "with the result that," and "in order that" are wordy and can be expressed more concisely ("also," "as a result") or dropped.

7-6 DELETE NEEDLESS WORDS OR PHRASES, BUT AVOID SHORTCUTS THAT SACRIFICE MEANING.

This is an editorial task—and a vital aspect of writing. We usually think of editing as what someone else does to our manuscript, but it is also what we ourselves do—a job that is much harder because the words are our own.

It is hard to edit ourselves because we assume that we have said what we tried to say. That is our first mistake. Our second mistake is that *we don't read the words.* The message that is recorded in our head is often not in the words themselves. And if it isn't, the reader will not get it. So first, we must assume nothing—that is, we must adopt the reader's perspective. And second, we must read the words—aloud, if necessary, so that the mind will not simply play back a prerecorded message. Only then are we ready to edit: to make those changes that will clarify meaning.

Of course, one writer's necessity may be another's trivia, or vice versa. Editing, like every task in writing, requires the exercise of judgment. In technical writing, the editorial standards are clarity and meaning. When in doubt, then, ask the question: Does this word (or phrase) clarify the intended meaning or merely create drag on the sentence? Usually the answer is surprisingly clear-cut. The commonly used phrases below, for example, contribute nothing to meaning; they either can be reduced to one word or dropped:

along the lines of
as to how long
by virtue of
for the purpose of
for the reason that
in accordance with
in connection with
in the absence of
in the event that
in relation to
insofar as
notwithstanding the fact that
the question of whether
the reason. . .was because
to the extent that
with a view to
with reference to
with regard to
with respect to
with the result that

Another drag on a sentence is *redundancy,* a special kind of repetition. A redundant word needlessly repeats the idea of something without repeating the word for it—for example, *hopeful optimism.* Sometimes it is less obvious, as in the sentence "Brown was thinking of possibly resigning," where *thinking* implies the idea of possibility, making *possibly* redundant.

Some commonly used redundant phrases are:

absolutely essential
actual experience
adequate enough
assemble together
cancel out
close proximity
consensus of opinion
each and every
few in number
first and foremost
free and clear
important essentials
minute droplets
mutual cooperation
null and void
one and the same
small in size
spherical in shape
through and through
uniformly consistent
yellow in color

7-7 CHOOSE CLARITY OVER STYLE IF THEY CONFLICT.

The objective, of course, is to write with both clarity *and* style. You can most nearly achieve this with lean, direct sentences, while keeping in mind the wholeness of the paragraph, the section, the report. The paragraph below was written by Loren Eiseley, a scientist whose writing is noted for its blend of clarity and style.

> Season followed season. Men grubbed for roots and nuts and berries in the short, cool summers, and followed remorselessly upon the trail of animals heavy with young. The spear and cutting flints were their only weapons; the bow was as far away in the future as gunpowder from the bow. All killing was personal. If a thrust failed, men stood to the charge of big game with nothing but a wooden spear and a flint knife. In the camps were children to be fed—big-browed children, the sides of whose foreheads were beginning to roof out with the heavy bone of their strange fathers.[1]

Note the simple but specific words (*roots, nuts, berries, wooden spears, flint knives,* and so on), the deliberate repetition of *followed,* and the variety in the length and structure of the sentences.

EXERCISES

Read the following paragraphs—

> Committee chairmanships are awarded through the controversial seniority system, in which uninterrupted length of service in the House or Senate is decisive. Though the system has been challenged, it still prevails. Those who defend the seniority system maintain that it supplies experienced, levelheaded leadership. The system's critics say it is unrepresentative and tends to entrench in power a disproportionate number of conservatives.
>
> The majority of American voters, the critics point

[1]Loren Eiseley, *The Firmament of Time* (New York: Atheneum, 1960), p. 106.

> out, are under 40; yet in the Senate 10 of the 16 chairpersons are past 65 and 6 are past 70; and in the House 11 of the 20 chairpersons are past 65 and 7 are past 70. Most of these chairpersons come from rural districts, which tend to be more conservative than urban districts.

1. Assume that the paragraphs describe the past. Revise all the verb tenses to reflect that change.

2. Identify the transitional device used to connect the two paragraphs.

3. Assume you are writing this description of the committee system in Congress, and the third paragraph will describe the assignment of members of Congress to committees. The facts are these: (1) Members of Congress request specific assignments; (2) these requests are based on preferences; (3) they also are based on experience; (4) they usually are awarded on the basis of seniority; (5) they sometimes are awarded as a favor; (6) they sometimes are awarded on the basis of influence; (7) they sometimes are denied as punishment. Using these facts, write a first sentence for the third paragraph, a sentence that eases the transition between the second and third paragraphs and even harks back to the first paragraph.

4. Rewrite the following sentence in a lean, more direct style:

> Being that the subject of death in our society is carefully avoided, terminally ill patients are not prepared to face it as a consequence.

5. In rewriting the following sentence, be as direct as possible and use only the essential information:

> This report summarizes the present state of my progress in the Technician Development program that is designed to offer guidance and training for advancement from an assembly line worker to a trained technician.

Despite its mental demands, writing is in the end a physical act. We must sit down and transform our thoughts into tiny marks on a blank sheet of paper. The distance between our head and that paper, less than a foot, can seem greater than a light-year.

Writing is hard work. But part of its difficulty stems from the way it is approached. The process of writing, for example, can be broken down into four stages: (1) planning, (2) researching, (3) the writing itself, and (4) revising. Most writers skimp on planning, researching, and revising, and focus on the writing itself. But research is important; it adds substance to what you write. And planning and revising are critical: solid planning makes the actual writing easier, and careful revising and rewriting make it more effective.

In writing, then, the key to method is the redistribution of the writer's time and attention, as discussed in this section. One suggested allotment: 40 percent for planning and research, 30 percent for the writing itself, and 30 percent for revising. Individual allotments will vary of course. But if the actual writing is allotted much more than one-third of the available time and attention, it will probably show the effects of insufficient planning, research, and revision.

8-1 PLAN A REPORT OR PAPER THOROUGHLY BEFORE STARTING TO WRITE IT.

If the planning is thorough, the writing will be easier. The first step is to ask yourself: What do I want to say? If you cannot answer that question, go no further. Though you may struggle bravely, what you produce will be minimally effective. Knowing what you want to say is like walking on firm ground instead of on quicksand, or like exploring with a compass instead of without one.

The next step in writing is to determine the purpose of the piece and the audience for it. Is it meant to be informative only, or persuasive or instructive as well? What do you know about the subject? About the audience? What does the audience know about the subject? What does the audience want? What does it want it for? And when does it want it?

Table 8-1 shows the results of an in-depth survey of Westinghouse managers who were asked what they looked for and what they wanted to know in the technical reports they read. The survey showed, for example, that all managers read the abstract or summary of a report, about half read the introduction and conclusion, and only a few read the body. Their responses are listed under the five categories that developed as the replies were tabulated. These are: (1) Problems, (2) New Projects and Products, (3) Tests and Experiments, (4) Materials and Processes, and (5) Field Troubles and Special Design Problems. Although the survey was published in 1962, it remains valid today—perhaps even more so in view of the current clamor in business and industry for more effective writing.

Once you know what the audience wants, the next step is to decide what you will say. Jot down all your thoughts about the subject on a piece of paper.

The next step is to organize those thoughts. Look for relationships between them—similarities, differences, cause-and-effect connections, before-and-after sequences. You can quickly connect related thoughts with lines to produce a rough outline. **(See also 4-2, p. 64.)**

TABLE 8-1. WHAT MANAGERS WANT TO KNOW

PROBLEMS
What is it?
Why undertaken?
Magnitude and importance?
What is being done? By whom?
Approaches used?
Thorough and complete?
Suggested solution? Best? Consider others?
What now?
Who does it?
Time factors?

NEW PROJECTS AND PRODUCTS
Potential?
Risks?
Scope of application?
Commercial implications?
Competition?
Importance to company?
More work to be done? Any problem?
Required manpower, facilities, and equipment?
Relative importance to other projects or products?
Life of project or product line?
Effect on company's technical position?
Priorities required?
Proposed schedule?
Target date?

TESTS AND EXPERIMENTS
What tested or investigated?
Why? How?
What did it show?
Better ways?
Conclusions? Recommendations?
Implications to company?

MATERIALS AND PROCESSES
Properties, characteristics, capabilities? Limitations?
Use requirements and environment?
Areas and scope of application?
Cost factors?
Availability and sources?
What else will do it?
Problems in using?
Significance of application to company?

FIELD TROUBLES AND SPECIAL DESIGN PROBLEMS
Specific equipment involved?
What trouble developed? Any trouble history?
How much involved?
Responsibility? Others? Company?
What is needed?
Special requirements and environment?
Who does it? Time factors?
Most practical solution? Recommended action?
Suggested product design changes?

Courtesy of the Westinghouse Electric Corporation

A rough outline not only tells you what you know, it also makes you aware of what you *don't* know. That is its greatest value. Look for consistency, logical sequence, and relevance. What is missing? What information is needed to complete this comparison?

What step is missing from that process? What is the history behind this policy? In technical writing, what you don't know can hurt you.

The next step is to research to fill in the gaps.

8-2 GATHER THE NECESSARY DATA: BASIC LIBRARY RESEARCH.

Any collection of books for use rather than for sale constitutes a library. It may be large or small, or a company, academic, or public library. The following, however, applies to large collections, which usually are found in academic or public libraries.

A large library has almost anything you need; the trick is to find it. The best source for help, of course, is a librarian. But research is easier when you know how a library is organized to retrieve its materials from the shelves. What follows, then, is: (1) a description of the basic retrieval systems; (2) a selective list of government publications; (3) a selective list of additional reference books; and (4) suggestions for taking notes.

Basic Retrieval Systems

The primary retrieval system is the card catalog. It lists all books and all other directories in the library—the indexes for periodicals, abstracts, newspapers, collections, dissertations, films, and more.

The card catalog consists of 3-by-5 cards filed alphabetically. Some libraries file alphabetically word by word, others file letter by letter. The difference is illustrated below:

WORD BY WORD	LETTER BY LETTER
New York	Newark
Newark	New York

The library materials themselves are organized under either of two classification systems: the Dewey Decimal or the Library of Congress. Generally, public libraries use the Dewey Decimal system and academic libraries use the Library of Congress system. Both systems employ numbers and letters to classify materials. But the principal dividers in the Dewey Decimal system are numbers; in the Library of Congress system the dividers are the letters of the alphabet.

The Dewey Decimal system. This system, devised in 1876 by Melvil Dewey, divides all knowledge into 10 basic subject classes, as shown in Table 8-2.

TABLE 8-2. THE 10 BASIC SUBJECT CLASSES IN THE DEWEY DECIMAL CLASSIFICATION SYSTEM.

000	General Works (encyclopedias, collections, indexes, and so on)
100	Philosophy and Related Fields (psychology, logic, ethics, and so on)
200	Religion
300	The Social Sciences (economics, education, law, and so on)
400	Language
500	Pure Science (astronomy, biology, physics, and so on)
600	Technology and Applied Sciences (engineering, medicine, agriculture, and so on)
700	The Arts
800	Literature
900	Geography and History

Each subject class is subdivided into 10 divisions, which are then further subdivided into 10 subdivisions. Tables 8-3 and 8-4 show the divisions of the 600 class, "Technology and Applied Sciences," and the subdivisions of the 620 division, "Engineering and Allied Operations."

TABLE 8-3. THE 10 DIVISIONS OF THE 600 CLASS, "TECHNOLOGY AND APPLIED SCIENCES," IN THE DEWEY DECIMAL CLASSIFICATION SYSTEM

600	Technology and applied sciences
610	Medical sciences
620	Engineering and allied operations
630	Agriculture and related technologies
640	Domestic arts and sciences
650	Managerial services
660	Chemical and related technologies
670	Manufactures
680	Miscellaneous manufactures
690	Buildings

TABLE 8-4.
THE SUBDIVISIONS OF THE 620 DIVISION,
"ENGINEERING AND ALLIED OPERATIONS,"
IN THE DEWEY DECIMAL CLASSIFICATION SYSTEM

620	Engineering and allied operations
621	Applied physics
622	Mining engineering and related operations
623	Military and nautical engineering
624	Civil engineering
625	Engineering of railroads, roads, highways
626	Unassigned*
627	Hydraulic engineering
628	Sanitary and municipal engineering
629	Other branches of engineering

*Unassigned subdivisions are reserved for future expansion.

Within these subdivisions, books are classified further by a book or author number placed after the subject class. The class number plus the book or author number is known as the *call number*. For example, the call number for Freud's *Civilization and Its Discontents* is 150.19, where 150 is the class/division (Philosophy/Psychology) and 19 is the book or author number. In many libraries the call number also includes the first letter of the author's name (150.19F); but the trend is to omit this letter. All books and materials are arranged on the shelves according to their call numbers.

The Library of Congress system. This system, created by Congress in 1897, was designed for large, expanding collections. Where the Dewey Decimal system, with its base of 10, provides 100 divisions, the Library of Congress system, based on the 26 letters of the alphabet, provides 676 divisions—the number 26 squared. The Library of Congress system also contains more basic subject classes. Table 8-5 shows these. Note that the letters *I, O, W, X,* and *Y* are not used; they have been reserved for future expansion.

TABLE 8-5. THE 20 BASIC SUBJECT CLASSES IN THE LIBRARY OF CONGRESS CLASSIFICATION SYSTEM

A	General Works (encyclopedias, indexes, and so on)
B	Philosophy, Psychology, Religion
C	Auxiliary Sciences of History (archaeology, biography, and so on)
D	History: General and Old World
E-F	History: American
G	Geography, Anthropology, Recreation
H	Social Sciences
J	Political Science
K	Law
L	Education
M	Music, Books on Music
N	Fine Arts
P	Language and Literature
Q	Science
R	Medicine
S	Agriculture
T	Technology
U	Military Science
V	Naval Science
Z	Bibliography, Library Science

The basic subject classes are subdivided by adding another letter, and these divisions are further subdivided by adding numbers. Division is according to need, not according to a consistent pattern, as in the Dewey Decimal system. Tables 8-6 and 8-7 on page 162, show the divisions of the T class, "Technology," and the subdivisions of the division TA, "Engineering. Civil Engineering." Those letters not used in Table 8-6 and the numbers not used in Table 8-7 have been reserved for future expansion. This, of course, is the main advantage of the Library of Congress system.

TABLE 8-6. THE 17 DIVISIONS OF THE T CLASS, "TECHNOLOGY," IN THE LIBRARY OF CONGRESS CLASSIFICATION SYSTEM

T	Technology
TA	Engineering. Civil engineering
TC	Hydraulic engineering
TD	Environmental technology. Sanitary engineering
TE	Highway engineering. Roads and pavements
TF	Railroad engineering and operation, including street railways and subways
TG	Bridge engineering
TH	Building construction
TJ	Mechanical engineering and machinery
TK	Electrical engineering. Electronics. Nuclear engineering
TL	Motor vehicles. Aeronautics. Astronautics
TN	Mining engineering. Metallurgy, including the mineral industries.
TP	Chemical technology
TR	Photography
TS	Manufactures
TT	Handicrafts. Arts and Crafts
TX	Home economics

TABLE 8-7. THE SUBDIVISIONS OF THE TA DIVISION, "ENGINEERING. CIVIL ENGINEERING," IN THE LIBRARY OF CONGRESS CLASSIFICATION SYSTEM

TA	Engineering. Civil engineering
TA 166–167	Human engineering
TA 168	Systems engineering
TA 177.4-185	Engineering economy
TA 349-360	Mechanics of engineering. Applied mechanics
TA 401-492	Materials of engineering and construction, including strength of materials
TA 501-625	Surveying
TA 630-695	Structural engineering (General)
TA 705–710	Engineering geology. Rock mechanics. Soil mechanics
TA 800–820	Tunneling. Tunnels
TA 1001-1280	Transportation engineering (General)
TA 1501-1820	Applied optics. Lasers, including applied holography, optical data processing
TA 2001-2030	Plasma engineering. Applied plasma dynamics

To these class numbers is added an author number, resulting in a call number. The author number includes the first letter of the author's last name. For example, the call number for Freud's *Civilization and Its Discontents* is BF173.F682, where BF is the class/division (Philosophy/Psychology) and F682 is the author number. All books and materials are arranged on the shelves according to their call numbers.

With the help of the card catalog, you can find a book if you know either the author or the title, or even if you know only the subject. If the library has the book, it will have a catalog card under all three headings. Figure 8-1 shows the author catalog card of Freud's *Civilization and Its Discontents.* The various notations on the card are numbered and explained below:

FIGURE 8-1. A TYPICAL LIBRARY CATALOG CARD

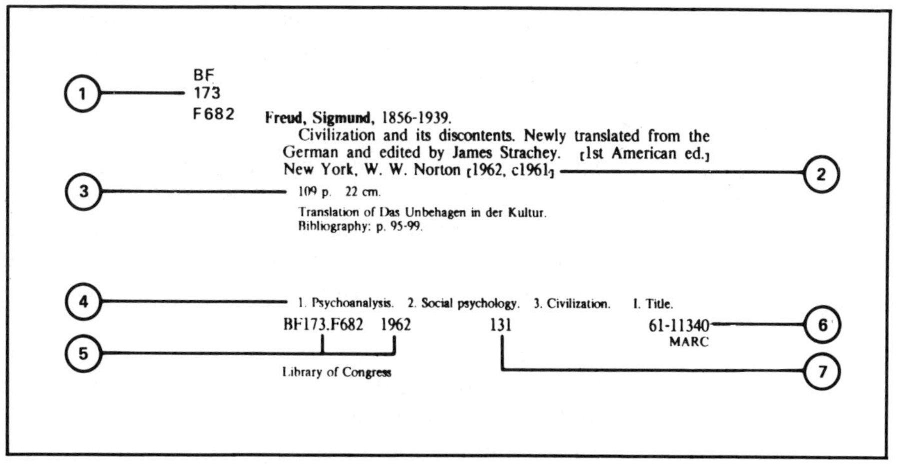

(1) the call number, here a Library of Congress classification; (2) the date of publication is 1962, the date of copyright, 1961; (3) the collation, which here includes the number of pages and the height of the book in centimeters, but also would mention illustrations, if the book contained them, and any series the book was a part of; (4) descriptions of the main subjects of the book, called "tracings" by librarians because they also are the headings under which the book is cross-indexed in the SUBJECT file of the catalog; (5) the Library of Congress classification number and the publication date, repeated here on the line librarians use for reference; (6) the Dewey Decimal classification number—which would also appear in the upper left-hand corner if the library used the Dewey Decimal instead of the Library of Congress system; (7) the Library of Congress catalog card number.

The most difficult research task is to find material when you have only a specific subject—no authors, no titles, nothing. In such a case, you need a *bibliography* —a list of sources of information—on your subject. Three basic sources of sources are listed on page 164:

Besterman's *A World Bibliography of Bibliographies* (4th ed., 1966).
Author, title, subject, and library index of 117,000 items grouped under 16,000 listings.

Bibliographic Index (1937-).
Subject index of bibliographies that are published separately or included in books or in the nearly 2000 periodicals examined.

Sheehy's *Guide to Reference Books* (9th ed., 1976).
Subject, title, and author index of 8000 sources in all fields. Divided into five sections: General Reference Works, the Humanities, Social Sciences, History and Area Studies, and Pure and Applied Sciences. Sources include books, periodicals, abstracts, dissertations, and more. (Sometimes referred to as Winchell's *Guide to Reference Books,* after the editor of previous editions.)

Another tool for either opening up or narrowing down your subject is the two-volume *Library of Congress Subject Headings* (8th ed., 1975).[1] Called the "Big Red Book" by librarians, it reflects the breakdown of subjects into their various and related aspects in the SUBJECT file of the card catalog.

For material in periodicals, consult the general indexes listed below; their location in a library is shown on their listing in the card catalog. Indexes are similar to bibliographies; the difference is that indexes are issued serially, that is, regularly, and so are more up to date than bibliographies. Most indexes are issued monthly, bimonthly, or quarterly, with a cumulative volume published annually.

When researching material in periodicals, begin with the latest articles and work backward. The most recent articles are more accurate and often contain bibliographies of earlier work. But before using an index—or any reference book—read its preface or introduction; it explains how to use the volume and provides a key to abbreviations and symbols used.

[1]Some libraries use instead the *Sears List of Subject Headings* (10th ed., 1972), which is based on the Library of Congress subject headings.

Applied Science and Technology Index (1913-).
Subject index to about 225 English language scientific and technical magazines. (Published as Part I of *Industrial Arts Index* from 1913 to 1957; Part II was *Business Periodicals Index.)*

Biography Index (1947-).
Name and profession index to biographical material in 1500 periodicals and in books. Covers the full range of biographical material, from genealogies to obituaries. The most comprehensive reference in the field.

Bolton's *Catalogue of Scientific and Technical Periodicals: 1665-1895* (2nd ed., 1965).
Titles of more than 86,000 scientific magazines published throughout the world since the rise of scientific periodicals.

Book Review Digest (1905-).
Reviews of books from about 75 English and American general magazines. Arranged alphabetically by author of book; also subject-title index.

Book Review Index (1965-).
Author index to reviews of books in more than 200 general magazines.

Business Periodicals Index (1958-).
Subject index to about 200 magazines covering every aspect of business and industry. (Published as Part II of *Industrial Arts Index* from 1913 to 1957; Part I was *Applied Science and Technology Index.)*

Pandex (1969-).
Subject, title, and author index to current scientific and technical publications, including books, periodicals, and reports. Also available on microform and on magnetic tape.

Reader's Guide to Periodical Literature (1900-).
Subject and author index to more than 200 general-interest magazines.

Technical Book Review Index (1935-).
Reviews from scientific, technical, and trade periodicals. Subject arrangement in the monthly issues; author index only in the annual collection. Since 1973, coverage includes life sciences, mathematics,

management, and behavioral sciences. (Also published during the years 1917 to 1928.)

Ulrich's *International Periodical Directory* (1932-).
Subject and title index to more than 57,000 scientific and technical journals. The most comprehensive index of scientific (especially scholarly) publications. Also contains an extensive list of abstract and indexing services.

Vertical File Index (1935-).
Subject and title index to selected pamphlets and booklets. (Published as *Vertical File Service Catalog* from 1935 to 1954.)

World List of Scientific Periodicals: 1900-1960 (4th ed., 1965).
Titles of more than 60,000 magazines on the natural sciences published throughout the world, nearly a quarter of them during the period 1951 to 1960.

A selected list of periodical indexes to publications in specific subject areas follows, arranged alphabetically. Their location in a library is also shown in the card catalog. For additional specific periodical indexes, consult Ulrich's *International Periodical Directory.*

Bibliography of Agriculture (1942-).
Author index (individual and corporate) to international list of periodicals, including lists of publications of the U.S. Department of Agriculture, state experimental stations and extension services, and the United Nations.

Biological & Agricultural Index (1964-).
Subject index to more than 200 periodicals, *excluding* publications of the U.S. and state governments and university research facilities. (Published as *Agricultural Index* from 1916 to 1964.)

Chemical Titles (1960-).
Author and title computer-produced index to 700 periodicals covering pure and applied chemistry and chemical engineering.

Current Index to Journals in Education (1969-).
Subject and author index to more than 700 periodi-

cals. Published in cooperation with the Education Resources Information Center (ERIC). For a guide to educational material dating back to 1929, see *Education Index*. (See also *Resources in Education*, p. 171.)

Engineering Index (1906-).
Subject and (after 1928) author index to about 1500 professional and trade periodicals and publications of societies, associations, universities, and government agencies throughout the world. Conference papers and selected books also covered. (Also published from 1892 to 1906, covering the period from 1884 to 1905.)

Humanities Index (1974-).
Subject and author index to about 260 periodicals covering history, language and literature, criticism and the performing arts, philosophy, religion, folklore, and related subjects. (Published as part of *International Index* from 1916 to 1964 and as part of *Social Sciences and Humanities Index* from 1965 to 1974.)

Index Medicus (1960-).
Subject and author index to about 2300 medical and health periodicals published throughout the world. Also includes selected periodicals in biometry, botany, entomology, psychology, sociology, veterinary medicine, and zoology. (Published under the titles *Index Medicus* from 1879 to 1927, *Quarterly Cumulative Index Medicus* from 1927 to 1956, and *Current List of Medical Literature* from 1956 to 1959.)

Index to Legal Periodicals (1908-).
Subject and author index to more than 400 periodicals. Since 1926, also a table of cases, and since 1940, a book review index as well.

International Nursing Index (1966-).
Subject index to 200 nursing journals and to articles about nursing in 2200 related periodicals indexed in *Index Medicus*.

Public Affairs Information Service Bulletin (1915-).
Subject index to more than 1000 English language periodicals, also to books, pamphlets, and reports of public and private agencies, covering economic, political, and social affairs. (Usually referred to as *PAIS Bulletin*.)

Science Citation Index (1961-).
Computer-produced index to the sources of articles on science and technology, medicine, agriculture, and the behavioral sciences from about 2000 periodicals. Classified by subject, citation of source, patent, author, and corporation. Also available on magnetic tape. (Not published in 1962 and 1963.)

Social Sciences Citation Index (1972-).
Computer-produced index to the sources of articles on social science from about 2000 periodicals. Classified by subject, citation of source, author, and corporation.

Social Sciences Index (1974-).
Subject and author index to about 270 periodicals covering anthropology, environmental science, geography, law, medicine, political science, psychology, and related subjects. (Published as part of *International Index* from 1916 to 1964 and as part of *Social Sciences and Humanities Index* from 1965 to 1974.)

A selected list of abstract journals follows, arranged alphabetically. Abstract journals are like indexes but include an abstract, or summary, of the indexed material. Like indexes, they are usually issued monthly, bimonthly, or quarterly, with a cumulative volume published annually. They also can be located in the card catalog, and an extensive list of them can be found in Ulrich's *International Periodical Directory.*

Abstracts in Anthropology (1970-).
Subject and author index (since 1971) to abstracts in archaeology, ethnology, linguistics, and physical anthropology. Covers books and conference papers as well as periodicals.

Applied Mechanics Reviews (1948-).
Author index to critical reviews of the contents of 1000 periodicals and hundreds of books published throughout the world. Divided into "Rational Mechanics and Math Methods," "Automatic Control," "Mechanics of Solids," "Mechanics of Fluids," "Thermal Sciences," and "Combined Fields." Annual

cumulative volume includes an author-title index, a list of reviewers, and a list of indexed publications.

Astronomy and Astrophysics Abstracts (1969-).
Subject and author index to abstracts from periodicals throughout the world, printed in English, French, and German. (Published as *Astronomischer Jahresbericht* from 1900 to 1969.)

Biological Abstracts (1926-).
Subject and author computer-based BASIC (Biological Abstracts Subjects in Context) index to abstracts from more than 5000 periodicals of 90 countries. Also a list of new books and periodicals. The most comprehensive reference to research in theoretical and applied biology.

Chemical Abstracts (1907-).
Subject and author index to abstracts from more than 14,000 periodicals in 50 languages. Additional indexes introduced since 1967 include those for patent numbers, key words, chemical substances, and general substances. The most comprehensive reference to research in chemistry. Also available on magnetic tape.

Computer & Control Abstracts (1966-).
See *Science Abstracts* below.

Electrical & Electronics Abstracts (1898-).
See *Science Abstracts* below.

International Aerospace Abstracts (1961-).
Semiannual subject and author index to abstracts of material published in periodicals and books. Also contract and report number indexes. (See also *Scientific and Technical Aerospace Reports* [*STAR*], p. 171.)

Mathematical Reviews (1940-).
Subject and author index to abstracts of the literature of pure and applied mathematics published throughout the world.

Metals Abstracts (1968-). Author index to abstracts from 1000 periodicals throughout the world, covering all aspects of metallurgy. (Published as *Metallurgical Abstracts* from 1934 to 1967.)

Oceanic Abstracts (1972-).
Subject and author index to abstracts of literature covering aspects of biology, geology, oceanography, pollution, and marine engineering. (Published as *Oceanic Index* from 1964 to 1967 and as *Oceanic Citation Journal* from 1968 to 1971.)

Physics Abstracts (1898-).
See *Science Abstracts* below.

Psychological Abstracts (1927-).
Author and category index to abstracts of material from periodicals, books, and reports. Since 1963, subject index also. *(Psychological Index* [1894-1935] lists earlier literature.)

Science Abstracts: Section A, *Physics* (1898-); Section B, *Electrical & Electronics* (1898-); Section C, *Computer & Control* (1966-).
Subject indexes to abstracts from periodicals, books, conference reports, and dissertations throughout the world. Author indexes and lists of periodicals added to semiannual issues. Available on magnetic tape and in microfiche editions.

For material in newspapers, consult the indexes below. Their location in a library also is shown on their listing in the card catalog.

New York Times Index (1913-).
Subject index to coverage by the *Times;* includes date, paper, column, and brief summary of each entry. Copies of the newspaper on microfilm. Index retroactive to 1851. The most comprehensive source of newspaper stories.

Newspaper Index (1972-).
Subject and name index to four major regional newspapers: the *Chicago Tribune, Los Angeles Times, New Orleans Times-Picayune,* and *Washington Post.*

Wall Street Journal Index (1958-).
Subject and corporation index to coverage of the *Journal,* which includes business news and national

affairs. Index in two sections: corporate news and general news.

For guides to special collections, dissertations, films, and other sources, see the card catalog or a librarian.

Government Publications

Government publications alone could fill an entire library. The following list of general and specific guides is necessarily selective; for additional titles, consult Jackson's *Subject Guide* or the *Monthly Catalog,* both listed below. All these publications can be located through the card catalog.

Congressional Information Service (1970-).
Subject and name index to abstracts of Congressional documents, including committee hearings and reports. Also bill, report, and document number indexes. (Usually referred to as *CIS Index.*)

E. Jackson's *Subject Guide to Major U.S. Government Publications* (1968).
Subject index plus a directory by W. A. Katz called "Guides, Catalogs and Indexes."

Monthly Catalog of U.S. Government Publications (1895-).
Agency and subject and (since 1974) title and author index. The most comprehensive list of government publications.

Resources in Education (1975-).
Subject, author, and institution index to abstracts of research reports in education. Published in cooperation with the Educational Resources Information Center (ERIC) of the U.S. Office of Education. (Published as *Research in Education* from 1966 to 1974. See also *Current Index to Journals in Education,* p. 166.)

Scientific and Technical Aerospace Reports (1963-).
Subject and author index to scientific and technical abstracts of reports issued by NASA, other govern-

ment agencies, and corporations and universities throughout the world. Also corporation and report/accession number indexes. Usually referred to as *STAR*. (See also *International Aerospace Abstracts,* p. 169.)

Statistical Abstract of the U.S. (1878-).
Covers the political, social, educational, commercial, and economic barometers of American life. Published by the U.S. Bureau of the Census.

UNDEX: United Nations Document Index (1970-).
Three indexes—Series A: Subject index; Series B: Country index; Series C (1974-): List of documents issued. Series A and Series B indexes computer-based. Information printed in English, French, Russian, and Spanish. Usually referred to as UNDEX. (*United Nations Document Index* also published from 1950 to 1973; superseded by UNDEX in 1974.)

U.S. Government Manual (1935-).
Official guide to the federal departments, agencies, bureaus, and so on. Lists activities and key personnel. Includes lists of former agencies and government publications.

U.S. Government Reports Announcements & Index (1971-).
Subject and author index to unclassified reports in science and industry. Also corporation and report/accession number indexes. Produced by the Department of Commerce. (Published as *U.S. Government Research Reports* from 1946 to 1965 and as *U.S. Government Research and Development Reports Index* from 1965 to 1971.)

Additional Reference Books

The following reference books are useful for finding specific information or for leads to additional sources of information.

Ayer Directory of Publications (1880-).
Title index of newspapers and periodicals published in the U.S., Canada, the Virgin Islands, Bermuda,

Panama, and the Philippines. Includes commercial data about states, counties, and cities. Arranged geographically.

Books in Print (1948-).
Subject, author, and title index of books currently in print. Also lists prices of books. See *Subject Guide to Books in Print* (1957-) for a list of subject headings and cross-references.

Business Books in Print (1973-).
Subject, author, and title index to books on business published in the U.S.

Dictionary of American Biography (1946).
Biographical sketches arranged by name of subject. Separate volume includes indexes by name, contributor, birthplace, schools attended, occupation, and topics. All subjects died before 1940. (For subjects living after 1940, see *Current Biography* [1940-]. Name and occupation index to biographical sketches of about 350 internationally prominent subjects annually.)

Encyclopedia of Business Information Sources (1970).
Lists bibliographies, periodicals, directories, and handbooks on (1) general and (2) geographical subjects of interest to business executives. Arranged alphabetically.

Gebbie Press House Magazine Directory (1952-).
Title index to house organs, that is, company magazines. Arranged alphabetically by name of company. Also geographical breakdown.

Paperbound Books in Print (1955-).
Subject, author, and title index to paperbacks currently in print. Also lists prices.

Thomas' Register of American Manufacturers (1905-).
Lists company names, addresses, telephone numbers, branch offices, and officials. Arranged alphabetically. Also index to products and services, and catalogs of companies.

Who's Who in America (1899-).
Biographical data on about 75,000 notable Americans, arranged by name of subject. See *Who Was Who*

in America, 5 vols. (1897-1973), for the late notables and *Who's Who of American Women* (1958-) for notable living women. Volumes also published for four U.S. regions, some states, and the world at large.

World Almanac and Book of Facts (1868-). Alphabetical index to miscellaneous information in just about every sphere of human activity.

Yearbook of the United Nations (1947-). Subject and name index to UN actions. Documentary bibliography included.

Taking Notes

Taking notes is an art and, as in all arts, the techniques of artists differ. Whatever technique used, however, notes should be selective yet complete—and concise.

Selectivity begins with the source itself. Is it relevant? Does it have the information you need? Check the table of contents or index of the source, or quickly scan the material itself.

If the source has what you need, what do you note down? It depends on what you are looking for, of course—background knowledge, a statistic, an example, a quote, a concept. Whatever it is, write it down in your own words on an index card or in a notebook, or on whatever works for you. Index cards can be shuffled around if you reorganize the piece, but they are easily lost. Notebooks are less easily lost, but they offer less flexibility.

What you write down should be complete enough to avoid a time-consuming return to the source. Delicate judgments of what you need are often involved here, and no writer judges correctly all the time. Some trips back to sources are inevitable, but keep them to a minimum. If your quarry is a portion of a quote, write down the entire quote so you will be able to blend it into your text skillfully. (Be sure, also, to use quote marks to indicate it is a quote.) If you are seeking background, take more rather than less so you won't miss a subtle but necessary element.

If a bibliography is involved, record all necessary

bibliographical information **(see section 6-8, p. 124)** on a separate card or notesheet, with a code (for example, a separate letter) for each source. Then add that code, with the page number, to each note from that source. The system is illustrated in Figures 8-2 and 8-3.

FIGURE 8-2.
A BIBLIOGRAPHICAL NOTESHEET

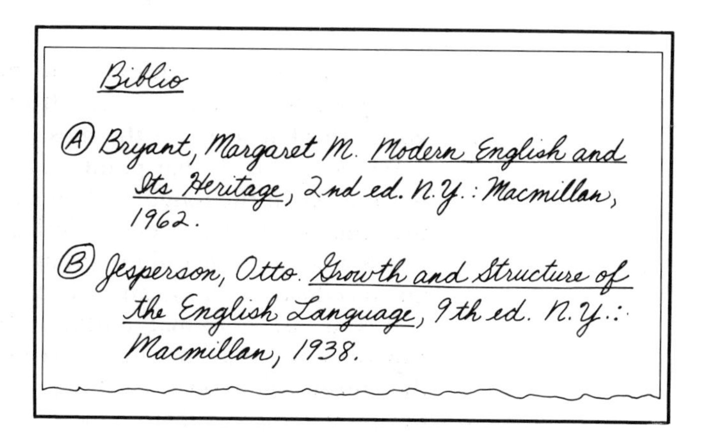

FIGURE 8-3.
A SAMPLE NOTE CARD

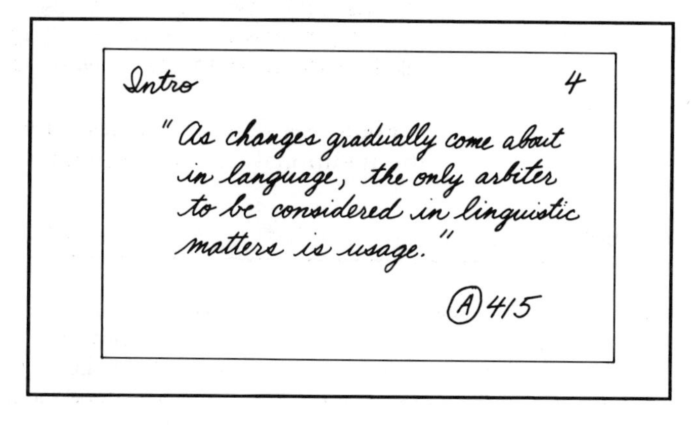

On Figure 8-3, "Intro" in the upper left-hand corner of the note card stands for "Introduction," the section of the report or paper where this information will be used. The number "4" in the upper right-hand corner indicates the relative position in the introduction where this item (among several others) might fit.

Conciseness is a desirable quality in all technical writing; in taking notes it is also a necessary one. Write down key ideas, key words, key phrases. (This also applies to notes of lectures or interviews.) Except when quoting, use a telegraphic style—mostly nouns and verbs—but do not omit necessary data.

A final word about research: it should be thorough. You probably will never use all of it, but you must *know* all of it to make the most of what you use.

When you reach that point, put together a working outline. As in the rough outline made before researching, aim for consistency, logical sequence, and relevance. Look for order of importance. Think in terms of an introduction, body, and conclusion.

If you use a formal outline, all entries must be consistent in phrase or sentence structure. For example, the entries below on the left are inconsistent; those on the right are consistent:

I. The use of memos	I. Using memos
II. Using numerals	II. Using numerals
III. How to use footnotes	III. Using footnotes

Figure 8-4 shows two types of formal outline formats, the traditional and the decimal:

FIGURE 8-4. FORMAL OUTLINES: TRADITIONAL AND DECIMAL FORMATS

I. FIRST CLASS	1. FIRST CLASS
A. First division	1.1 First division
1. First subdivision	1.1.1 First subdivision
2. Second subdivision	1.1.2 Second subdivision
a. First sub-subdivision	1.1.2.1. First sub-subdivision
b. Second sub-subdivision	1.1.2.2 Second sub-subdivison
B. Second division	1.2 Second division
II. SECOND CLASS	2. SECOND CLASS
A. First division	2.1 First division
B. Second division	2.2 Second division
C. Third division	2.3 Third division

8-3 WRITE A FIRST DRAFT.

Here the advice of the King of Hearts to the White Rabbit in *Alice in Wonderland* is pertinent: "Begin at the beginning, and go on till you reach the end; then stop."

Begin by stating the purpose of the piece; continue by developing what you have to say; and end by tying everything together in a conclusion or summary. Write to the shape of your material. Simple material lends itself to straight exposition. But complex material may demand a weaving together of several strands of thought so that related themes appear, fade, then reappear in different contexts. The difference is like that between a halfback who bursts through the opposing defense and races straight down the field to score, and one who weaves through the defense, giving a little here, taking away some there, until reaching the end zone.

Don't be afraid to deviate from your outline as you write. The shape of a piece, scientist Gregory Bateson reminds us, "sometimes emerges out of a sort of wrestling process." So rather than hang on to a plan that no longer reflects reality, go with the flow of the piece as it develops. Most of us learn as we write—one of the prime benefits of writing. **(See also 4-1, p. 63.)**

8-4 REVISE AND REWRITE—AS OFTEN AS NECESSARY.

Editing your own writing requires you to see yourself as others see you—a difficult and sometimes painful task. But you can do it if you shift from your own perspective to that of the reader and *read the words themselves* rather than simply rethink the intentions in your head. What do the *words* mean? Have you said what you wanted to say? Is it clear? Could it be clearer? More concise?

Unless you are a professional editor, read your work aloud. Your ears may pick up what your untrained eyes miss, especially typos and careless repetition and omissions.

After revising, write a second draft and edit that. Repeat the process as often as necessary. The writing that is clearest and most readable is almost always the writing that has been revised and rewritten most. As in so much else, you get out of it what you put into it.

EXERCISES

1. Focus on a specific aspect of one of the subjects listed below (for example, containing forest fires in remote wilderness areas) and select a title for a projected report on the subject:

air pollution	heart transplants
alcoholism in children	herbicides
behaviorism	holograms
biofeedback	legalization of marijuana
cholesterol	man-made satellites
criminal law	radioactive wastes
cryogenics	sailing
Darwin	Saturn probe
divorce	solar heating
DNA	sound systems
Euclid	taxation
fire fighting	transistors
Freud	tunneling
glaciation	UFOs
	water pollution

2. Prepare a bibliography of at least 10 titles for the projected report. The bibliography should include periodicals as well as books.
3. Examine one of the periodicals and write a one-page memo describing its size and layout as well as content. Discuss its scope, style, and value as a source.
4. Write a similar memo describing one of the books.
5. Using the materials above, make a formal outline for the projected report.

APPENDIX 1. ANALYSIS OF A SAMPLE REPORT

The memo in Figure A-1 is an example of a clear, concise presentation of specific information about a complex subject. In this case it is the function, financing, governing, and managing of California community colleges in the post-Proposition 13 era, when the colleges' former income from community property taxes has been sharply reduced. It could, however, be any subject. For the technical writer, the problem is always the same: how to communicate the necessary information accurately and effectively.

In terms of the audience and purpose, the organization and presentation of information, and the tone and length of the memo, note the following points:

1. The memo is addressed to the SRJC staff, and its purpose is to inform them of the specific factors that will affect the future function, financing, governing, and managing of California community colleges—a subject of great concern to the staff.

2. Mikalson organizes the 20 specific factors according to the two committees that will deal with them: Governor Brown's Commission on Government Reform and Chancellor Craig's Task Force for Community College Finance. This introduces the memo. The description of how the Task Force will operate comprises the body of the memo. Mikalson then concludes by speculating about the Task Force's success and by spelling out what he will do and what further information is available to the staff.

3. The subject of the memo is precisely stated under SUBJECT. The rest of the memo is planned (thought out or outlined) before a word of it is written.

4. In the first paragraph Mikalson immediately places the problem in context by citing the goals of the Commission—whose recommendations, of course, will affect what happens to community colleges. Then he lays out the specific problems the colleges must deal with. The numbering of factors in both paragraphs adds clarity.

5. The word *Meanwhile* establishes a relationship between the first and second paragraphs, making the

FIGURE A-1. A SAMPLE MEMO REPORT

September 20, 1978

TO: ALL STAFF
SANTA ROSA JUNIOR COLLEGE

FROM: ROY G. MIKALSON
SUPERINTENDENT/PRESIDENT

SUBJECT: COMMUNITY COLLEGE PLANNING FOR
1979-1980 AND BEYOND

Shortly after the passage of Proposition 13, Governor Brown appointed his Commission on Government Reform, chaired by A. Alan Post, to:

1. develop a formula for distributing the one (1) per cent property tax;

2. investigate local governmental functions appropriate for transfer to state jurisdiction;

3. study the impact of Proposition 13 on public employment;

4. forecast State revenues;

5. recommend appropriate tax and revenue sharing plans; and,

6. strengthen long-term revenue expenditure balances.

The commission will make an interim report to the Governor and legislature by January 15, 1979, on the financing and structure of State and local government.

Meanwhile, Chancellor Bill Craig has appointed a Task Force for Community College Finance to study and report back in four (4) general areas:

1. missions and functions of community colleges;
2. finance;
3. governance; and,
4. management.

The sub-topics are:

1. program content and access;
2. program priorities;
3. support sources;

4. support techniques;
5. decision-making structure;
6. district and college organization;
7. inter-institutional coordination;
8. use of college resources;
9. scheduling and measuring student work; and,
10. accountability.

Study groups will be meeting in each area, studying alternative approaches, discussing and recommending action or courses of action to the main group. The Task Force will be meeting with varied groups throughout the state to get input from local colleges and lay people and will be receiving input from any interested people or groups. The Chancellor has assigned the equivalent of seven (7) full-time staff to work on the project.

Once options or alternatives are developed by the smaller groups, the Task Force will begin a careful analysis of all options and will run simulations to see how districts are affected. The Task Force will give a report to the Board of Governors in October and November and will present a finance package to the Board of Governors in December or early January so legislation can be prepared for July 1, 1979 implementation.

The overall hope is that a long-term package, which covers not only the financing but the function, governance and control of community colleges, can be approved by the Board of Governors, and then by the legislature. To accomplish this, the package will probably have to have consensus approval from the field and most associations. Should this fail, we shall probably have another one-year interim package put together by the Department of Finance or the legislature.

During the next month I shall be at two or three meetings at which the project will be thoroughly discussed. I shall write regular reports on them and let everyone know not only what is happening, but also how one can get input into the process.

The Task Force is made up of members of the Chancellor's staff; representatives from the California Postsecondary Commission, the Public Employment Relations Board, State Department of Finance, legislative staff people; private citizens; faculty members; administrators; and students. If anyone wishes to see the list of people on this Task Force, a copy is in my office.

RGM:lb

first more relevant. Without this transition, the implication that the Commission's recommendations will affect what happens to community colleges could be lost on the reader.

6. In the fifth paragraph Mikalson sums up the problem. He then suggests what will be required for approval of an effective long-term package and speculates on what the alternative will be, failing approval.

7. In the last two paragraphs Mikalson states what he will do and offers the reader information that will be useful if the reader wishes to act. A case can be made for placing the last paragraph before the third one. The composition of the Task Force logically belongs there. But the offer to the reader of the names of the Task Force members—the implied call to action, so to speak—belongs at the end of the memo, where its impact is strongest.

8. Although the characteristic response to the passage of Proposition 13 has been emotional, the tone of the memo is dispassionate and professional.

9. The memo is only two pages long, a limit Mikalson says he tries not to exceed.

APPENDIX 2. OVERVIEW OF GRAMMAR

For writers, the striking fact about grammar is that the study of it does not necessarily improve writing ability. Despite many attempts, researchers in the field have been unable to establish a cause-and-effect connection between the study of grammar and improvement in writing. What connection there is seems negative: as more time is spent studying grammar, less is spent writing—and the writing improves less.

Yet many writers have problems with sentence structure, tense, the function of words, and so on—in short, with the subject matter of grammar. So some knowledge of grammar seems essential for them. The question is, *what* knowledge? And the answer to it seems clear: the knowledge essential to solve their specific writing problems. For example, how can I write a complete sentence? What is the difference between a phrase and a clause? When does a word like *writing* function as a verb, when as a noun, and when as a modifier? And so on.

The basic question behind these questions is: how do words function in English? What follows examines the basic functions of the basic parts of the sentence.

HOW WORDS FUNCTION

As noted on page 33, words can be divided into two categories: *meaning* words and *structure* words. *Meaning* words convey the basic meaning of a sentence; *structure* words help hold a sentence together, add meaning to it, and refine its style.

There are only four types of meaning words: nouns and verbs (the core of a sentence) and adjectives and adverbs (often simply called modifiers). All other words—pronouns, prepositions, conjunctions, and so on—are structure words. In *this sentence,* for *example, all* the *meaning words are* in *italics.*

Figure A-2, on page 186, shows the relationships of the four types of meaning words and what each one does in the following sentence:

ADVERB	ADJECTIVE	NOUN	VERB	ADVERB	ADVERB
Extremely	hard	metal	lasts	much	longer.

FIGURE A-2.
THE RELATIONSHIPS OF THE FOUR TYPES OF MEANING WORDS AND HOW EACH ONE FUNCTIONS

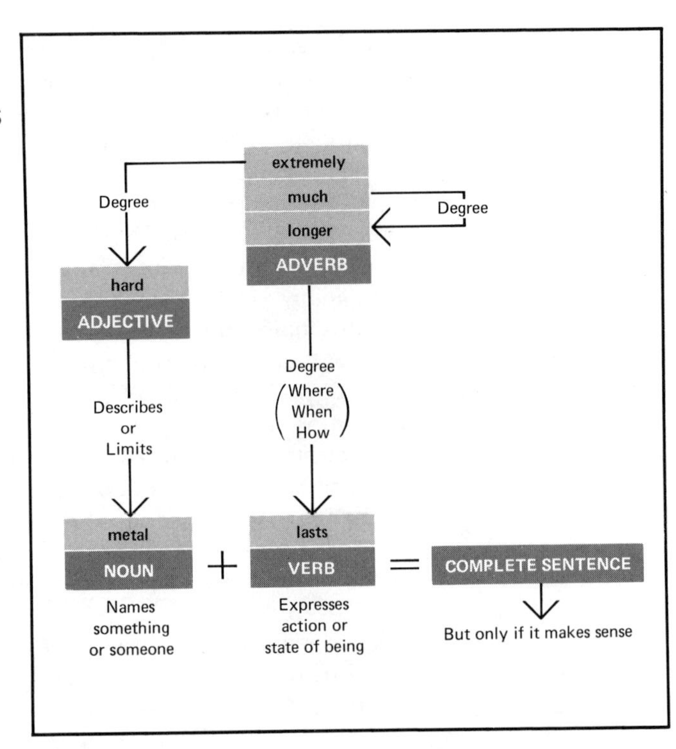

Note that all the adverbs in Figure A-2—*longer, extremely,* and *much*—indicate degree in the sentence. However adverbs that modify verbs also can indicate place, time, and manner. For example, in the sentence "Goodman drove fast and arrived there early," *fast, there,* and *early* are adverbs. *Fast* indicates how Goodman drove; *there* indicates where Goodman arrived; and *early* indicates when.

Similarly, the adjective *hard* in Figure A-2 describes the noun *metal,* but an adjective may also specify limits, as in the sentence "*Ten* houses were built this week."

VERBS AND VERBALS

Verbal is the name given to the *infinitive, present participle,* and *past participle* of a verb when they do not function as a verb.

Verbals usually function as a modifier, sometimes as a

noun. But it is easy to mistake a verbal for a verb because the forms are nearly identical. For example, the infinitive of the verb *learn* is *to learn,* the present participle is *learning,* and the past participle is *learned.* Only a slight difference distinguishes the forms of the present tense, *learn,* and the present progressive, *am learning,* from those of the verbal infinitive, *to learn,* and the present participle verbal, *learning.* And there is no difference between the form of the past tense, *learned,* and that of the past participle verbal, *learned.*

To distinguish between a verb and verbal, determine how they function in a sentence: whether they act as a verb, or as a modifier or noun. Table A-1 shows how the three verbals and their comparable verb forms function in a sentence:

TABLE A-1. THE FORMS AND FUNCTIONS OF VERBALS

FORM	FUNCTION	EXAMPLE OF USE IN A SENTENCE
Infinitive	Verb Adjective	not applicable The president of the company is a good person to know.
	Noun	"To write is to sit in judgment on one-self"—Ibsen.
	Adverb	The Federal Reserve acts to control the supply of money.
Present Participle	Verb	Americans are reading less and watching TV more.
	Adjective Noun*	Every writing job is a thinking job. Technical writing is first of all writing.
Past Participle	Verb	We have not yet learned to reduce our expectations.
	Adjective	Civil service applicants must take a written exam.

*When the present participle functions as a noun, it is also called a gerund.

HOW UNITS OF LANGUAGE FUNCTION

Words, the basic unit of writing, combine to form larger elements that function as single units in a sentence.

These larger elements are phrases and clauses.

Phrases and clauses are examined in detail in Section 3, pages 35 to 37. Here the emphasis is on their function as a unit. The distinctive difference, remember, between a phrase and a clause is that a clause contains a subject and predicate, and a phrase does not.

Phrases. A phrase (depending on what kind it is) usually functions as a modifier (an adjective or adverb), sometimes as a noun, and only in one instance (that of a verb phrase) as a verb. In the following examples, the phrases are underlined and their function is noted:

Tomorrow the company opens a new office
MODIFIER
in Des Moines.

NOUN
A number of witnesses described the accident differently.

MODIFIER VERB
By July 31 all the inventory will have been sold.

Note that *in Des Moines* tells us *where* the new office will be opened; that *A number of witnesses* is the subject of the sentence; and that *By July 31* tells us *when* the inventory will have been sold.

Clauses. An independent clause does not function as anything in a sentence because it stands complete in itself and *is* a sentence. But a dependent clause (which depends on an independent clause to complete its meaning) may function as a modifier or noun. In the following examples, the clauses are underlined and their function is noted:

NOUN
Most economists predict that the economy will get worse before it gets better.

MODIFIER
When imports rise, the value of the dollar decreases.

Note that the dependent clause *that the economy will get worse before it gets better* tells us *what* the economists predict, and that the dependent clause *When imports rise* tells us *when* the value of the dollar decreases.

Table A-2 shows how the various types of phrases and the two types of clauses may function in a sentence. For example, the table shows that a prepositional phrase always functions as a modifier, a noun phrase always functions as a noun, and a verbal phrase or a dependent clause may function as either a modifier or noun.

TABLE A-2. FUNCTIONS OF PHRASES AND CLAUSES IN SENTENCES

PHRASES						
Function as	Prep	Verbal			Noun	Verb
		Infinitive	Participle	Gerund		
Adjective modifier	Yes	Yes	Yes	No	No	No
Adverb modifier	Yes	Yes	No	No	No	No
Noun	No	Yes	No	Yes	Always	No
Verb	No	No	No	No	No	Always

CLAUSES		
Function as	Dependent	Independent
Adjective modifier	Yes	Not applicable
Adverb modifier	Yes	Not applicable
Noun	Yes	Not applicable
Verb	Not applicable	Not applicable

APPENDIX 3. OVERVIEW OF SPELLING

Like ornamental ironwork, spelling is increasingly a lost art, and the trend is likely to continue. It is easy to see why, for English spelling is both irrational and inconsistent. The 26 letters of the alphabet symbolize at least 46 sounds—and these are spelled 456 different ways! For example, the *ain* sound alone has 6 spellings: c*ane,* r*ain,* r*ein,* r*eign,* camp*aign,* and champ*agne.* And the spelling combination *ough* has at least 8 different sounds, including *cough, bough, tough, through,* and *thorough.*

Even so, of the 20,000 or so commonly used words, only about 1 percent are consistently misspelled. And of these 200, the dozen that cause the most trouble are:

believe	disappoint	occasion
coolly	embarrass	occurrence
definitely	existence	receive
disappear	losing	separate

To those dozen add the following dozen, words that look and sound like other words but are spelled differently—and have different meanings:

to affect them	their product	then turn left
the effect on them	over there	more than one
its color	they're leaving	to New York
it's traditional		too high
		two tickets

Consistently poor spellers probably need a special course in spelling. But writers with only isolated blindspots may be able to eliminate them with some memorization and the will to look up doubtful words in a dictionary. Most spelling errors can be avoided by following the IE, the final E, and the doubling guidelines, and by memorizing the exceptions to each. The guidelines and the exceptions are listed below:

THE IE

When the sound is long *e* (as in *believe*), put *i* before *e* except after *c*: *achieve, chief, field, piece,* and so on.

In addition, use *i* before *e* when they follow a *sh* sound: *ancient, conscience, deficient, species,* and so on.

The following 10 words still take *i* before *e* even though the guidelines above do not apply to them:

audience	hierarchy
fiery	medieval
financier	notoriety
friend	quiet
gaiety	society

The following 10 words are *exceptions* to the guidelines; they take *e* before *i:*

deity	leisure
either	neither
foreign	seige
forfeit	sheik
height	weird

THE FINAL E

When a word ends in *e,* drop the *e* before a suffix beginning with a vowel—but keep the *e* before a suffix beginning with a consonant: *hoping, hopeful; caring, careful; using, useful;* and so on.

The following 13 exceptions retain the *e* before a vowel suffix:

acreage	noticeable
changeable	outrageous
courageous	replaceable
dyeing[1]	serviceable
knowledgeable	singeing[2]
manageable	traceable
mileage	

The following 7 exceptions drop the *e* before a consonant suffix:

[1]To distinguish it from dying.

[2]To distinguish it from singing.

argument
awful
duly
ninth
truly
wholly
wisdom

THE DOUBLING

If the final syllable of a word is accented (com·pel′) and it ends in a single consonant preceded by a single vowel, double the final consonant before a suffix beginning with a vowel: *compelled, controlled, patrolled, preferred,* and so on.

But, if the accent is *not* on the last syllable (ben′·e·fit) do *not* double: *benefited, canceled, counselor, traveled,* and so on. (These words, it should be noted, are usually considered correct with or without the doubled consonant.)

The following exception does not double, even though the accent is on the last syllable of the base word:

chagrined

The following two exceptions do not double because the accent, which is on the last syllable in the base word, shifts to the first syllable when the suffix is added:

conference
preference

The following five exceptions double, even though the accent is not on the last syllable of the base word:

crystallize
legionnaire
metallic
questionnaire
tranquillity

One final guideline: memorize what you can and resolve any doubts with a dictionary.

APPENDIX 4. CONVERSION TABLES

The metric system is based on the decimal system—that is, on the number 10. Its basic units are: (1) the *meter,* for length; (2) the *liter,* for volume; (3) the *gram,* for weight; and (4) degrees *Celsius,* or *centigrade,* for temperature.

Three prefixes are used with the basic metric units:

milli = one-thousandth (0.001)

centi = one-hundredth (0.01)

kilo = one thousand (1000)

For example: 1000 millimeters = 1 meter; 100 centimeters = 1 meter; 1 kilometer = 1000 meters.

Figure A-3 shows how the Fahrenheit and Celsius, or centigrade, temperature scales compare; and Tables A-3 and A-4, on pages 196 and 197, show how the basic metric and U.S. units of measure compare. For example, Table A-3 shows that 1 kilometer = 0.621 miles, and Table A-4 shows that 1 mile = 1.609 kilometers. The tables also can be used to convert units of measure from one system to the other.

FIGURE A-3 A COMPARISON OF FAHRENHEIT AND CELSIUS TEMPERATURE SCALES

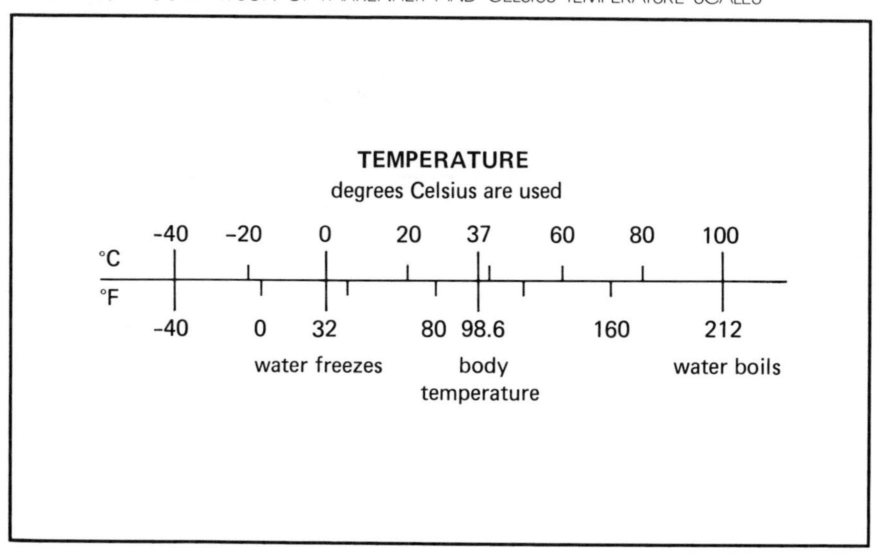

TABLE A-3.
CONVERSION TABLE:
U.S. TO METRIC

TO CONVERT FROM U.S. TO METRIC:		MULTIPLY THE U.S. UNIT BY:
length		
inches	millimeters	25.400
inches	centimeters	2.540
inches	meters	.025
feet	meters	0.305
yards	meters	0.914
miles	kilometers	1.609
area and volume		
cubic inches	cubic centi-meters	16.387
square inches	square centi-meters	6.451
square feet	square meters	.093
square yards	square meters	0.836
liquid measure		
fluid ounces	milliliters	29.573
U.S. quarts	liters	0.946
U.S. gallons	liters	3.785
cubic feet	liters	28.339
cubic inches	liters	.016
weight and mass		
grains	grams	.065
ounces	grams	28.349
pounds	kilograms	0.453

All numbers rounded to three decimals.

TABLE A-4.
CONVERSION TABLE: METRIC TO U.S.

TO CONVERT FROM METRIC TO U.S.		MULTIPLY THE METRIC UNIT BY:
length		
millimeters	inches	.039
centimeters	inches	0.394
meters	inches	39.370
meters	feet	3.280
meters	yards	1.093
kilometers	miles	0.621
area and volume		
cubic centimeters	cubic inches	.061
square centimeters	square inches	0.155
square meters	square feet	10.764
square meters	square yards	1.196
liquid measure		
milliliters	fluid ounces	0.034
liters	U.S. quarts	1.057
liters	U.S. gallons	0.264
liters	cubic feet	.035
liters	cubic inches	61.020
weight and mass		
grams	grains	15.430
grams	ounces	.035
kilograms	pounds	2.205

All numbers rounded to three decimals.

INDEX

Page numbers in *italics* refer to illustrations; those in **boldface** refer to definitions. Numbers followed by the letter n (124n) are references to footnotes.

Abbreviations, 85, 125–29
Abstract, 96, **110**, *111*. *See also* Formal report
Abstract words, 19
Accident report, 7. *See also* Reports
Accuracy, 2, 6, **10**, 26
Acronyms, 85
Active voice, 50, 51. *See also* Verbs
Adjective, **22**, 33
Adverb, **22**, 23, 33
  conjunctive, **40**, *41*, 42, 79
Agreement. *See* Nouns; Pronouns; Verbs
Alice in Wonderland, 24, 177
Alliteration, 149
Alphabet, 191
Ambiguity, 10, 22, 23
American English, 28, 32, 71
American Psychological Association, 52
Analogies, **146**, 147
Analysis, 6, 15, 55, 62, **63**–65, 156. *See also* Logical analysis
Analytic language, 31
Anglo-Saxons, 49. *See also* Words
Annual report, 6. *See also* Reports
Antecedents, 46
Antonyms, 27. *See also* Thesaurus; Words
Any, 43, 45
Apostrophe, 47, 72, **81**
Appendix of a report, 96. *See also* Formal report
Appositive, 87
Approvals list, formal report, 95. *See also* Formal report
Approximate numbers, 123
Assumption in editing, 151
Asterisk, 125
Audience, 3, 7, **9**, 10, 55, 95–97, 156, *157*, 180
Authority, 3, 16
Auxiliary verbs, 36, 143, 146. *See also* Verbs

Balance, 62
Bar, 72, **74**. *See also* Slash
Basic retrieval systems. *See* Dewey Decimal system; Library of Congress system
Bateson, Gregory, 90, 177
Bibliography, 96, 124
  of basic sources, 163–74
  for formal reports, 96, 97
  preparation of, 131–36, 174, *175*
  for this book, 135
Blueprint, 122
Body, of report or letter, 2, 66–68, 93, 98, 101, 108, 177, 180. *See also* Conclusion; Introduction; Letters; Organization; Paragraph
Books in Print, 135
Brackets, 72, **89**, 89n
Bridgeman, P. W., 51
Built-up sentence, 139–41. *See also* Lean sentence

Caesar, Julius, 138
Case
  in nouns, 43n, 47
  in pronouns, 43n, 46
Causal relationships, 38, **64**, 65, 156
Cause-and-effect, 38, **64**, 65, 156
Celsius, 195
Charts. *See* Graphs
Christensen, Francis, 66–68, 139, 140
Clarity, 10, 12, 16, 18, 50, 141, 146–48, 151–53, 177
Classification, 63
Clauses
  as adjective, 189
  as adverb, 189
  as basic element of sentence, 35
  dependent, **36**, 37, 40, 46, 56, 188, 189
  independent, **36**, 37, 40, 188, 189
  main, 37
  as modifier, 37, 39, 188, *189*
  nonrestrictive, 77, 78
  as noun, 188, 189
  restrictive, 77, 78
  subordinate, 36, 56, 57, 67. *See also* Sentences
Clichés, **24**, 25, 27
Coherence, 10, **11**, 12
Collective nouns, 44

Colon, 72, 78, 80, 84, 90
Comma, 40-42, 71, 72, 74-81, 90
Comparisons, 11-16, 67, 115, **146**, 147
Complaint report, 6, 7. *See also* Reports
Complement of verbs, 37. *See also* Verbs
Complete sentence, 33, 40. *See also* Sentences
Complexity, 2
Compound subjects, 43-48. *See also* Verbs
Compound words. *See* Hyphen
Conciseness, 10, **12**, 13, 141, 151, 174, 176
Conclusion, of report or letter, 93, 98, 102, 108, 177, 180, 183. *See also* Body; Introduction; Letters; Organization; Paragraph
Conditional mood, 146. *See also* Verbs
Conference report, 7. *See also* Reports
Conjunction, 33, 40-42, 57, 75
  coordinating, 40, *41*, 42
  subordinating, *57*
Conjunctive adverb, **40**, 41, 42, 79
Connotative meaning, 20. *See also* Words
Consistency, 62, 63, 147, 157
  in use of numerals, 123, 124
Context, 40, 58
Contractions, *81*, *82*, 83
Contrasts, 115
Conversion tables, *195*, *196*, 197
Coordination, 56, 57, 66-68. *See also* Sentences; Subordination
Cover, formal report, 95. *See also* Formal report
Cumulative sentence, **139**, 141. *See also* Lean sentence
Cycles, 14

Dagger, 125
Dangling modifier, 39
Dash, 72, 78-80, 86-88, 90
Data sheet, 105, 106, *107*, 108. *See also* Letter report
Decimal system, 195
Deduction, 15, 65
Deletion, 151, 177
Demonstrative pronouns, **46**, 47. *See also* Pronouns
Denotative meaning, 20. *See also* Words
Dependent clause, **36**, 37, 40, 46, 56, 188, 189. *See also* Clauses
Description, 2, 13-16, 19, 94, 113, 141, 146
Development, 64, 66-68, 93, 176, 177, 180, 183. *See also* Body; Conclusion; Introduction; Letters; Organization; Paragraph
Dewey, Melvil, 159
Dewey Decimal system, 158, *159*, *160*, 161
Diagrams. *See also* Drawings; Graphs
  assembly plan, 122
  block, 118
  blueprint, 122
  of circuit, 118, *119*
  map, *122*, 123
  schematic, 118, *119*
Dialogue, 79, 80, 90, 91
Dictionaries, 27-29
Dictionary of Scientific and Technical Terms, 28
Directions, 7, *8-9*
Distribution list, formal report, 95. *See also* Formal report
Double dagger, 125
Drawings, 113, *121*, 122, 123. *See also* Diagrams; Graphs
  cutaway, 122
  exploded-view, *121*, 122

Editing, 151, 152, 155, 177-78
Einstein, Albert, 16, 142
Eiseley, Loren, 153
Ellipses, 72, 73, 74
Embedded phrases, 36. *See also* Phrases
Emerson, Ralph W., 74
Emphasis, 3, 50, 64, 149
"Et al.," 125
Euphemism, 10, 24, **25**, 26, 27
Examples, 13-15, 146, 147
Exclamation point, 71-73, 90
Explanation, 2, 16, 67
Exposition, 2, 16, 67

Fahrenheit, 195
Feasibility report, 6. *See also* Reports
Feasibility study, 6
Federal Standardization Handbook, 3
First draft, 176, 177
Focus, 13-16
Footnotes, 97, **124**, 124n, 125, 125n, 126-31

Foreword, formal report, 96. *See also* Formal report
Formal report, *95*. *See also* Reports; Technical writing
abstract of, 96, 110, *111*
appendix of, 96
approvals list of, 95
audience of, 95, 96, 97
bibliography of, 96, 97
cover, 95
distribution list of, 95
footnotes in, 97
foreword of, 96
glossary of, 96
index of, 97
length of, 95
letter of transmittal for, 96
list of illustrations in, 96
preface to, 96
purpose of, 97
summary of, 95, 110, *112*
table of contents of, 96
title page of, 95
titles in, 94
Format, **93**, 94–136. *See also* Bibliography; Diagram; Drawing; Footnotes; Formal report; Letter report; Reports
Form vs. meaning, 44, 45
Form vs. usage, 28, 31, 32, 44, 45, 48
Fragment sentence, 39, 40. *See also* Sentences
Frost, Robert, 46

Gender in pronouns, 47, 48, 52
Gerund, *187n*
Glossary, formal report, 96. *See also* Formal report
Gobbledygook, 11, 26
Gram, 195
Grammar, 31–33, 185–89
and writing, 185
Grammarians, prescriptive and descriptive, 31, 32, 48
Graphics, 18. *See also* Diagrams; Drawings; Graphs
Graphs, 113, *115*. *See also* Diagrams; Drawings
bar, 115, *116*
circle, 115, *117*
flow charts, 115, *118*
line, *115*, 116
organizational charts, 115, 118, *120*

Hall, Donald, 24, 138
Headword, 36, 45
Heisenberg, Werner, 51
Helping verb, 36, 143, 146. *See also* Verbs
Hyphen, 72, 85

"Ibid.," **128**
Ideal sentence, 141, 142. *See also* Sentences
Ideas, number of in a sentence, 56. *See also* Sentences
Illustrations. *See* Diagrams; Drawings; Graphs; Tables
Immediacy, 142, 143
Imperative mood, 146. *See also* Verbs
Incoherence, 12
Independent clause, **36**, 37, 40, 188, *189*. *See also* Clauses
Index, formal report, 97. *See also* Formal report
Indicative mood, 146. *See also* Verbs
Indirect question, 72–73
Induction, 15, 65
Infinitive, 35, 43, 186, *187*. *See also* Verbals; Verbs
Inflated language, 23–27. *See also* Words
Inflated sentence, 139, *141*. *See also* Rambling sentences; Sentences
Inflection, 31, 32, 43, 43n, 47. *See also* Words
Informal report, *95*, 97–109. *See also* Letter report; Memo report; Reports
Information, 2, 3
distortion of, 38
omission of, 38, 39
Instruction, 3, 7
sample work procedure instructions, *8*, *9*
Instruction manual, 3, 7
Introduction, of report or letter, 93, 98–101, 108, 177, 180. *See also* Body; Conclusion; Letters; Organization; Paragraph; Salutation
Investigative report, 7. *See also* Reports
Italics, 73

Janus, 149. *See also* Transition
Jargon, 10, 24, **26**, 27.

Job application letter, 105, 106, *107*, 108. *See also* Letters
Johnson, Samuel, 90

Kierkegaard, Soren, 90
King of Hearts, 177
Kozol, Jonathan, 91

Laboratory notebook, 7, 109. *See also* Laboratory Report
Laboratory report, 7, 109-10. *See also* Reports
Laing, R.D., 91
Language
  analytic, 31
  function of units of, 187-89
  inflated, 23-27
  inflected, 31, 32, 43, 43n, 47
  and the sentence, 31
  spoken and written, 18
  synthetic, 31
  vague, 19
Lean sentence, 139, 141, 142
Length of formal report, 95. *See also* Formal report
Length of sentence, 55, 56. *See also* Sentences
Letter of transmittal, 96. *See also* Formal report
Letter report. *See also* Reports
  adjustment, *104*
  body of, 66-68, 93, 99, 101, 108
  complaint, *103*
  conclusion of, 93, 98, *101*, 108
  format of, *99*
  inquiry, *102*
  introduction to, 93, 99-101, 108
  job application, 105, 106, *107*, 108
  resumé or vita or data sheet, 105, 106, *107*, 108
Letters, 97, 98, *99*-108. *See also* Job application letter
Library of Congress system, 158, 159, 160, *161*, *162*, *163*
Library research. *See* Research
Lincoln, Abraham, 76
List of illustrations, formal report, 96. *See also* Formal report
Liter, 195
"Loc. cit.," 128
Logical analysis, 6, 15, 55, 62-65, 156

Main verb, 36, 50. *See also* Verbs
Manual of Style, A, 135
Manuals, 2, 7, 8. *See also* Reports; Technical writing
Maps, *122*, 123. *See also* Diagrams
Marx, Karl, 90
Meaning words, 33, 185, *186*. *See also* Sentences; Structure words
Memo report, *4*, *5*, 108-09, 180, *181-82*, 183. *See also* Reports
Meter, 195
Method, **155**, 156-77
  planning a report, 156-58
  researching, 158-76
  revision and rewriting, 177
  writing first draft, 176-77
Metric system, 195, *196*, *197*
Middle English, 32
Mikalson, Roy, 180-83
Modern English, 32
Modifiers, 22, 23, **33**, 34, 35, 37, 39, 40, 188, *189*. *See also* Adjective; Adverb; Clauses; Phrases
Monotony, 34, 55, 58. *See also* Sentences
Mood, 146. *See also* Verbs

Necessity, 13, 59
Negative connotation, 20. *See also* Words
New American Roget's College Thesaurus, 29
None, 43, 44, 45
Notebook. *See* Laboratory notebook; Laboratory report; Notes
Notes. *See also* Footnotes; Laboratory notebook
  bibliography, 174, 175
  conciseness in taking, 176
  index card vs. notebooks, 174
  note cards, *175*
  note-taking, 174, *175*
  selectivity, 174
Noun phrase, 35, 36, 44, 45, *189*. *See also* Phrases
Nouns, **22**, 23

agreement with pronouns, 43, 46
case in, 43n, 47
collective, 44
function of, 22
as meaning word, 33, 185, *186*
number in, 43, 43n, 46, 47
strength of, 22
use of apostrophe with, 47
Nuances, 19
Number, 43, 43n, 46. *See also* Nouns; Pronouns; Verbs
Numbers
in addresses and highways, 124
approximate, 123
beginning a sentence, 123
with commas, 124
consistency in use of, 123
in decimals, 124
in fractions, 124
with hyphens, 123
in lists, 124
mixed, 124
under 10, 123
under 20, 123
in units of measurement, 124
used consecutively, 124
use of numerals, 123, 124
written-out, 123
Numerals. *See* Numbers

Object, description of, 13, 14
Objectivity, 2, 15, 51
Object of preposition, 35. *See also* Phrases
Object of verb, **34**, 50. *See also* Verbs
Obstructions to clarity, 10, 11. *See also* Language; Static verb phrases; Words
Old English, 32, 48
Only, 54
"Op. cit.," 128
Order, 62. *See also* Development; Organization; Outlines
Organization, **62**, 63-68, 180, 183. *See also* Development; Order; Outlines
analysis, 62, **63**, 147, 157, 176
basic procedures, 62, 63
basic unit of, 66
classification, **63**
consistency, 62, 63, 147, 157, 176
logical analysis, 6, 15, 55, 62, **63**-65, 156
of paragraph, 66-68
relevance, 13, 59, 62, 157, 174, 176, 180, 183
of report, 64, 66
sequence, 62, 64, 156, 157, 176
synthesis, *63*
visible aspects of, 62
Orwell, George, 26, 27
Outlines, 65, 66, 156, 157, *176*, 177. *See also* Development; Order; Organization
Oxford English Dictionary, 29

Papers. *See* Technical papers
Paragraph, 64, 66-68. *See also* Development; Order; Organization
as basic unit of organization, 66
coordination and subordination in, 67, 68
development of, 64, 66-68, 177
emphasis in, 64
length of, 67
organization, 66-68
topic sentence in, 66, 67
traditional approach to, 66, 67
transitions between, 66-68, 149, *150*, 151, 181, 183
Parallel structure, 58, 59
Pared-down sentence, 139, *141*, 142. *See also* Sentences
Parentheses, *72*, 78, 89, 90
Passive voice, 47-49, 50-53. *See also* Verbs
Past participle, 35, 50, 143, 186, *187*. *See also* Verbs
Period, 40-42, 71, *72*, 75, 90
Person
in pronouns, 43, 46
in verbs, 43n
Persuasion, 3, 6, 16
Phrases
adjective, *189*
adverb, *189*
deletion of, 151
as element of sentence, 35
embedded, 36
function as unit, 188, *189*
gerund, 36, *187*
headword of, **36**, 45
infinitive, *189*

as modifier, 35, 39, 40, *189*
needless, 151
noun, 35, 36, 44, 45, *189*
participial. *See* Verbals
prepositional, 35, 36, *189*
repetition of, 147
static verb phrases, 36n, 49. *See also* Verb phrases; Verbs
verb, 35, 36, 36n, *189*
verbal, 35, 39, 40, *189*
Policies, 13, 16-17
Positive connotations, 20. *See also* Words
Possessive, 46, 47
Precision, 2, 22, 94
Predicate adjective, **34**, 37
Predicate noun, **34**, 37
Predicate of sentence, 33, **34**, 36, 37. *See also* Sentences; Verbs
Preface, formal report, 96. *See also* Formal report
Prefixes, 85, *86*
Preposition, 33, 35, 36, 189
Prepositional phrases, 35, 36, *189*. *See also* Phrases
Present participle, 35, 143, 186, *187*. *See also* Verbs
Principles of technical writing, 1-17. *See also* Responsibilities of writer; Technical writing
Processes, 14, 15
Progressive form, 143. *See also* Verbs
Progress report, 6. *See also* Reports
Project report, 6. *See also* Reports
Pronouns, *46*
agreement with antecedents, 43, 46
case in, 43n, 46
demonstrative, 46, 47
gender in, 47, 48, 52
number in 43, 46
personal, 43, 46
person in, 46
relative, 46
sexism in, 47, 48, 52
as structure word, 33
Proposal, 3, *4*, *5*
Punctuation, **71**-91. *See also* Apostrophe; Brackets; Colon; Comma; Dash; Ellipses; Exclamation point; Hyphen; Parentheses; Period; Question mark; Quote marks; Semicolon; Slash
to enclose, *72*, 77-79, 86-89
to end, 72
function of, 18, 71, *72*
history of, 71
inconsistency in, 71
to introduce, *72*, 79-81, 84, 86, 88
to separate, *72*, 75-77, 84-87
tendency in, 71
uses of, 42, *72-74*
Purpose, 3, 55, 156, 180, 183

Question mark, 71, *72*, 73, 90
Quote marks, *72*, 74, 90, 91

Rambling sentences, 139, 141, 142. *See also* Inflated sentence; Sentences
Random House Dictionary of the English Language, 28
Reading level, 23, 24, **55**, 67. *See also* Audience; Vocabulary; Words
Redundancy, 152
Reed, James A., 91
Reference books. *See* Sources, reference
Reiteration. *See* Repetition
Relative pronouns, 46. *See also* Pronouns
Relativity, theory of, 16, 142
Relevance, 13, 59, 62, 157, 174, 176
Repetition, 147-51
Reports. *See also* Formal report; Letter report; Technical writing
accident, 7
accuracy in, 10
annual, 6
audience for, 9, 10
complaint, 6, 7
conference, 7
essential qualities of, 10-13
feasibility, 6
formal, 95
format of, 93, *95*, 110, 111
informal, *95*, 97-109
investigative, 7
laboratory, 7, 109-10
length of report, 95, 97, 180, 183
letter, 97, 98, *99*, 100-01, *102*, *103*, *104*, 105-108
manuals, 2, 7, 8
memo, *4*, *5*, 108-09, 180, *181-82*, 183

organization of, 64-66
progress, 6
project, 6
proposal, 3, *4*, *5*, 6
purpose of, 3-8
research, 7
status, 6
survey, *4*, *5*, 6
technical level of, 10. *See also* Audience; Words
titles of, 94
tone of, 3
trip, 7
trouble, 7
Research, 10, 155, 158-76
Research report, 7. *See also* Reports
Responsibilities of writer, 2, 10, 26, 97, 105, 113, 123, 176, 177, 180. *See also* Technical writing
Resumé, 105, 106, *107*, 108. *See also* Letter report
Revision, 151, 152, 155, 177-78
Rewriting, 151, 152, 155, 177-78
Roget, Peter Mark, 29
Roget's *Thesaurus*, 29
Rules vs. usage, 28, 31, 32, 44, 45, 48

Saint-Exupery, Antoine de, 13
Salutation, 99, 100. *See also* Introduction; Letters
Santayana, George, 90
Selectivity, 2, 139, 174
Semicolon, 40-42, 71, *72*, 75, 84, 85, 90
Sentences, **31**-59. *See also* Words
ambiguity in, 51
basic elements of, 33, 35-37
built-up, 139-41
causal relationships in, 38, **64**, 65, 156
complete, 33, 40
complex, 56, 57. *See also* Subordination
compound, 56, 57. *See also* Coordination
context of, 58
coordinate and subordinate, *56*, 57, 66-68
correct, 37-42
cumulative, **139**-41
dangling modifier in, 39, 40
deletion of, 59
distortion in, 38, 53
emphasis in, 50, 149
ending with preposition, 54
fragment, 39, 40
framework of, 34, 35
ideal, 141-42
ideas in, 56, 57. *See also* Coordination; Subordination
importance of position, 31, 53, 54
inflated, 139, *141*, 142
lean, 139, 141, 142
length of, 55, 56
meaning words in, 33, 185, *186*. *See also* Structure words
as measure of writing, 31
monotonous, 34, 55, 58
omissions in, 38-40
parallel structure in, 58, 59
pared-down, 139, 141, 142
patterns of, 31, 33, 37
predicate of, 33, 34, 36, 37
rambling, 139, *141*, 142. *See also* Inflated sentence
recasting of, 59
run-on, 39-42
shifting voices in, 52, 53. *See also* Active voice; Passive voice
simple, 55, 56
structure, 33, 53-56, 58, 59
structure words in, 33, 185. *See also* Meaning words
subject of, 33, 34, 36, 37, 39, 45, 50
topic, 66, 67. *See also* Paragraph
transition between, 149, *150*, 151. *See also* Paragraph
Sequence, 62, 64, 156, 157, 176
Shortcuts, 151
Significant detail, 19, 139. *See also* Specific words
Simplicity, 142-46. *See also* Style; Words
Slash, *72*, 74
Sources, reference
abstract journals, 168-70
government publications, 171, 172
Library of Congress Subject Headings, 164
for material in newspapers, 170, 171
for material in periodicals, 164-66
for specific information or leads, 171-74
for specific periodical indexes, 166-68
three basic, 164

Specifications
government, 3
military, 3
non-military, 3
Specifications, Types and Forms, 3
Specific words, 19, 31. *See also* Significant detail
Spelling, 27, 29, 191-93. *See also* Dictionaries; Words
Split infinitive, 54. *See also* Verbs
Standardization Policies, Procedures and Instructions, 3
Static verb phrases, 10, 36n, 49. *See also* Verb phrases; Verbs
Status report, 6. *See also* Reports
Stevens, Charles F., 15
Structure of sentences, 33, 53-56, 58, 59. *See also* Sentences
Structure words, 33, 185. *See also* Meaning words; Sentences
Style, **138**-53. *See also* Sentences; Words
basic dimension of, 138
change in, 138
clarity and, 146-48
comparison of, 138
direct, 139-42
essence of, 138
formal or informal, 7
function of, 138
inflated, 139, *141*, 142
lean, 139-41
selectivity in, 139. *See also* Sentences; Words
simplicity in, 142-46. *See also* Words
technical dimension of, 138. *See also* Technical writing
vs. clarity, 153
the writer and, 138
Subjunctive mood, 49, 146. *See also* Verbs
Subordinate clause, 35, 56, 57, 67. *See also* Clauses
Subordination, 56, 57, 66-68
Subverbo, 127
Suffixes, 85, *86*
Summary, formal report, 96, **110**, *112*. *See also* Formal report
Survey report, *4*, *5*, 6. *See also* Reports
Surveys, subjects of, 6
Swift, Jonathan, 27
Synonyms, 27-29, 147
Syntax, 31, 53, 54
Synthesis, *63*
Synthetic language, 31

Table of contents, formal report, 96. *See also* Formal report
Tables, 113, *114*
metric conversion, 195, *196*, *197*
Technical articles, 7. *See also* Reports; Technical writing
Technical level, 10. *See also* Audience; Words
Technical papers, 7. *See also* Reports; Technical writing
Technical report. *See* Reports
Technical writer
ability to perceive relationships, 63
choice of illustrations, 123
responsibilities of, 2, 10, 26, 97, 105, 113, 123, 176, 177, 180,
vocabulary, 29
Technical writing. *See also* Editing; Reports
audience for, 3, 7, **9**, 10, 55, 95, 96, 97, 156, *157*, 180
basic function of, 2, 3, 13-17. *See also* Purpose
bibliography, 96, 131-36, 174, 175
categories of, 3, 6, 7
essential qualities of, 10-13. *See also* Clarity; Coherence; Conciseness
footnotes in, 97, 124, 124n, 125, 125n, 126-31
grammar in, 185. *See also* Grammar
levels of, 10. *See also* Audience; Words
nature of, 1
organization of, 62, 63-68, 180, 183. *See also* Development; Order; Organization
origins of, 1
principles of, 1-17
priority of, 2
process of, 155-58, 176, 177. *See also* Method
purpose of, 3, 177
special knowledge in, 1, 2, 13
specific format in, 93
standard of, 2
subjects of, 13-16
tone of, 3, 10

Tenses, *143*. *See also* Verbs
action in, 142
agreement in, 43, 43n
changing, 146
differences in, 143
forms for, *143*
future, 143
future perfect, 143
imperfect, 143
need for auxiliary verbs, 50, 143
past, 142
past perfect, 143
perfect, 143
present, 142, 143
present perfect, 143
problems of overlap in, 144, 145
time covered by, 142, *145*
Theory, 15, 109
Thesaurus, 27, 29
Title page, formal report, 95. *See also* Formal report
Titles, 94. *See also* Formal report
Tone, 3, 10, 180, 183
Topic sentence, 66, 67. *See also* Paragraph; Sentences
Transitions, 149, **150**, 151, 181, 182, 183. *See also* Paragraph; Sentences
Trip report, 7. *See also* Reports
Trouble report, 7. *See also* Reports
Twain, Mark, 22

Uncertainty Principle (Heisenberg), 51
Usage, 32, 37, 44, 48
Usage vs. form, 28, 31, 32, 44, 45, 48

Verbal phrase, 35, 39, 40, *189*. *See also* Phrases
Verbals, forms and functions of, 35, 186, 187
Verb phrases, 10, 11, 35, **36**, 36n, 49, *189*. *See also* Verbs
Verbs, **22**, 23, 33. *See also* Tense; Verbals
active voice, 50, 51
agreement with subject, 43
auxiliary, 36, 146
complement of, 37
compound subject of, 43, 45
conditional mood of, 146
helping, 36, 146
imperative mood of, 146
indicative mood of, 146
infinitive of, 35, 43, 186, *187*. *See also* Verbals
main, 36, 50
as meaning words, 33
mood of, 146
number in, 43, 43n
object of, 34, 50
passive voice, 47-49, 50-53
past participle of, 35, 50, 143, 186, 187, *189*. *See also* Verbals
person in, 43n
phrases, 10, 11, 35, **36**, 36n, 49, *189*
present participle of, 35, 143, 186, *187*. *See also* Verbals
progressive form of, *143*
split infinitive of, 54
static verb phrases, 10, 36n, 49
strength of, 22, 49, 50
strong vs. weak, 36n, 49, 50
subject of, 45, 50
subjunctive mood of, 49, 146
tenses of, 43, 50, 187
use of singular or plural, 44, 45, 46
Vidal, Gore, 14
Virgule, *72*, **73**. *See also* Slash
Vita, 105, 106, *107*, 108. *See also* Data sheet; Letter report; Resumé
Vocabulary, 2, 18, 27, 29. *See also* Reading level; Words
Voice, 47-49, 50-53. *See also* Active voice; Passive voice; Verbs

Walden pond, 2
Webster's New International Dictionary of the English Language, 28
Webster's New World Dictionary of the American Language, 28
Webster's Third New International Dictionary, 1, 28, 93
White Rabbit, 177
Whose vs. *the . . . of which*, 48. *See also* Form vs. usage; Grammarians
Who vs. *whom*, 48. *See also* Form vs. usage; Grammarians
Wordiness, 10, 12, 13, 22, 23
Words, 18-29. *See also* Sentences; Vocabulary
abstract, 19

ambiguous, 10, 22, 23. *See also* Ambiguity
Anglo-Saxon, 18, 24
antonyms, 27. *See also* Thesaurus
arrangement of, 2, 138
basic unit of writing, 18, 35, 187
cliches, 24, 25
connotative meaning of, 20
deceptive, 21, 22, 24, 26, 81-83
definitions of, 20
deletion of, 151, 177
denotative meaning of, 20
element of sentence, 18, 35
euphemism, 10, 25, 26
forceful, 23, 24
French, Latin, and Greek, 18, 23, 24
function of, 18, 19, 35, 185, *186*, 187. *See also* Grammar; Sentences; Syntax
general, 19
Indo-European, 18
inflated, 23-27
inflection of, 31, 32, 43, 43n, 47, 80, 81
jargon, 10, 24, **26**, 27
lifeless, 24
long vs. short, 23, 24
meaning, 19, 33, 185, *186*
misused or misplaced, 10, 19-23, 53, 54
negative connotations, 20
obscure, 23, 24
order of, 31
plain vs. fancy, 23, 24
positive connotations, 20
pretentious, 23, 24
pronunciation of, 27, 29
relationship between, 31, 33, 35, 185, 186
repetition of, 147-51
roots of, 1, 29, 65
short vs. long, 23, 24
similar, 19-22
specific, 19, 31
spelling of, 27, 29, 191-93
structure, 33, 185
synonyms, 27-29, 147
ten most common, 18
vague, 10, 11, 19, 20. *See also* Ambiguity
Writer's responsibilities, 2, 10, 26, 97, 105, 113, 123, 176, 177, 180
Writing. *See also* Editing; Reports; Revisions; Technical writing
clear, 10, 12, 16, 18, 50, 141, 146-48, 151-53, 177
coherent, 10, **11**, 12
concise, 10, **12**, 13, 141, 151, 174, 176
planning of, 11, 155, 156, *157*, 158. *See also* Method
process of, 155, 176, 177. *See also* Method
revising of, 151, 152, 155, 177, 178
Writing process, 155-58, 176, 177. *See also* Method

A 0
B 1
C 2
D 3
E 4
F 5
G 6
H 7
I 8
J 9